Jacq Cobhar

100 MATHS HOMEWORK ACTIVITIES

D1325135

CONTENTS

**Published by
Scholastic Ltd,
Villiers House,
Clarendon Avenue,
Leamington Spa,
Warwickshire CV32 5PR**

© Scholastic Ltd 2001
Text © Sonia Tibbatts and John Davis 2001
Additional material on
pages 6–8 © Ann Montague-Smith 2001
2 3 4 5 6 7 8 9   2 3 4 5 6 7 8 9 0

**AUTHORS
Sonia Tibbatts and John Davis**

**EDITORIAL & DESIGN
Crystal Presentations Ltd**

**COVER DESIGN
Joy Monkhouse**

**COVER ARTWORK
Philippa Dally**

**ILLUSTRATOR
Phil Garner**

Acknowledgements

The publishers wish to thank:
Ann Montague-Smith for her invaluable advice in
the development of this series.
The Controller of HMSO and the DfEE for the use
of extracts from *The National Numeracy Strategy:
Framework for Teaching Mathematics* © March 1999,
Crown Copyright (1999, DfEE, Her Majesty's
Stationery Office).

British Library Cataloguing-in-Publication Data

A catalogue record of this book is available from the British
Library.

ISBN 0-439-01847-1

The right of Sonia Tibbatts and John Davis to be
identified as the Authors of this work has been asserted
by them in accordance with the Copyright, Designs and
Patents Act 1988.

PAGE IN THIS BOOK	ACTIVITY NAME	HOMEWORK	STRAND	TOPIC	NNS UNIT	LESSON	PAGE
29	Abacus charts	Practice exercise	Numbers and the number system	Place value	①	①	⑲ ⑳
30	Add or subtract	Maths to share	Numbers and the number system	Ordering and rounding	①	②	⑳ ㉑
31	Jumping kangaroo	Games and puzzles	Numbers and the number system	Number sequences	② ③	②	㉔ ㉕
32	Number splits	Practice exercise	Calculations	Mental calculations + and –	② ③	④	㉖
33	In a word	Games and puzzles	Calculations	Recall of + and – facts	② ③	⑦	㉘
34	All change	Maths to share	Solving problems	Number problems in money	② ③	⑧	㉘ ㉙
35	Odd and even	Games and puzzles	Numbers and the number system	Number properties	② ③	⑨	㉙ ㉚
36	Different order	Investigation	Calculations	Checking results	② ③	⑩	㉚
37	Measure up	Practice exercise	Measures, shape and space	Measures	④ ⑥	②	㉞ ㉟
38	Long distance	Investigation	Measures, shape and space	Measures	④ ⑥	④ ⑤	㉟ ㊱
39	Cover up	Games and puzzles	Measures, shape and space	Measures	④ ⑥	⑥ ⑦	㊲
40	All shapes and sizes	Maths to share	Measures, shape and space	Shape and space	④ ⑥	⑨ ⑩	㊳
41	In the net	Investigation	Measures, shape and space	Shape and space	④ ⑥	⑪ ⑫	㊴
42	Double trouble	Practice exercise	Calculations	Recall of × and ÷ facts	⑧	①	㊽ ㊾
43	Times three	Games and puzzles	Calculations	Recall of × and ÷ facts	⑧	② ③	㊾ ㊿
44	Through the maze	Maths to share	Calculations	Recall of × and ÷ facts	⑧	④	㊿ 51
45	Magic squares	Investigation	Solving problems	Reasoning about numbers	⑧	⑤	51 52
46	In the family	Practice exercise	Calculations	Understanding × and ÷	⑨ ⑩	①	54
47	Number trios	Maths to share	Calculations	Understanding × and ÷	⑨ ⑩	②	55
48	Double cross	Games and puzzles	Calculations	Mental calculations × and ÷	⑨ ⑩	④ ⑤	56 57
49	Top ten	Maths to share	Calculations	Mental calculations × and ÷	⑨ ⑩	⑥	57 58
50	On the grid	Practice exercise	Calculations	Paper and pencil × and ÷	⑨ ⑩	⑦	58 59
51	Left overs	Games and puzzles	Calculations	Understanding × and ÷	⑨ ⑩	⑧	59
52	Big spender	Investigation	Solving problems	Number problems in 'real life'	⑨ ⑩	⑨	60
53	Spot the fraction	Practice exercise	Numbers and the number system	Fractions	⑪	①	63 64

Column group headers: **100 MATHS HOMEWORK ACTIVITIES YEAR 4** · **NATIONAL NUMERACY STRATEGY** · **100 MATHS LESSONS**

REFERENCE GRID

100 MATHS HOMEWORK ACTIVITIES YEAR 4 / NATIONAL NUMERACY STRATEGY / 100 MATHS LESSONS

PAGE IN THIS BOOK	ACTIVITY NAME	HOMEWORK	STRAND	TOPIC	NNS UNIT	LESSON	PAGE
79	Find the rule	Maths to share	Numbers and the number system	Number sequences	8	1 / 2	112 / 113
80	Odds and evens	Games and puzzles	Numbers and the number system	Number properties	8	3	113 / 114
81	Number puzzles	Games and puzzles	Solving problems	Reasoning about numbers	8	4 / 5	114 / 115
82	Tables bingo	Maths to share	Calculations	Recall of × and ÷ facts	9 / 10	1	120
83	Rearrange it!	Practice exercise	Calculations	Understanding × and ÷	9 / 10	2	120 / 121
84	Using the grid!	Practice exercise	Calculations	Paper and pencil × and ÷	9 / 10	3 / 4	121 / 123
85	Approximate first!	Timed practice exercise	Calculations	Paper and pencil × and ÷	9 / 10	6 / 7	123 / 124
86	Check it!	Investigation	Calculations	Checking results	9 / 10	8	124 / 125
87	What's left?	Games and puzzles	Calculations	Understanding × and ÷	9 / 10	9	125 / 126
88	The school trip	Investigation	Solving problems	Making decisions	9 / 10	9	125 / 126
89	Fraction match	Timed practice exercise	Numbers and the number system	Fractions	11	1 / 2	129 / 130
90	Name the fraction!	Investigation	Numbers and the number system	Fractions	11	1 / 2	129 / 130
91	Make 1	Games and puzzles	Numbers and the number system	Fractions	11	3	130 / 131
92	Using fractions	Practice exercise	Numbers and the number system	Fractions	11	4 / 5	131
93	Pictogram	Practice exercise	Handling data	Organising data	12	1 / 2	132 / 133
94	Which books?	Maths to share	Handling data	Interpreting data	12	4 / 5	134
95	Graph it!	Investigation	Handling data	Organising data	12	4 / 5	134
96	Ton up	Games and puzzles	Numbers and the number system	Place value	1	1	143 / 144
97	Give me a sign	Practice exercise	Numbers and the number system	Ordering and rounding	1	2	144 / 145
98	Top tens	Maths to share	Calculations	Mental calculations + and −	2 / 3	1	148 / 149
99	Magic machines	Maths to share	Calculations	Mental calculations + and −	2 / 3	2	149 / 150
100	Column addition	Practice exercise	Calculations	Paper and pencil + and −	2 / 3	4 / 5	151 / 152
101	Column subtraction	Practice exercise	Calculations	Paper and pencil + and −	2 / 3	6 / 7	152
102	School stock	Investigation	Calculations	Paper and pencil + and −	2 / 3	8	152 / 153
103	Holiday travel	Games and puzzles	Solving problems	Number problems in 'real life'	2 / 3	9	153 / 154

100 MATHS HOMEWORK ACTIVITIES YEAR 4			NATIONAL NUMERACY STRATEGY		100 MATHS LESSONS		
PAGE IN THIS BOOK	ACTIVITY NAME	HOMEWORK	STRAND	TOPIC	NNS UNIT	LESSON	PAGE
104	Matching measures	Practice exercise	Measures, shape and space	Measures	4 6	1	160
105	Capacity quiz	Maths to share	Measures, shape and space	Measures	4 6	3 4	161 162
106	Mirror image	Games and puzzles	Measures, shape and space	Shape and space	4 6	6	163
107	Letter land	Maths to share	Measures, shape and space	Shape and space	4 6	7 8	163 164
108	Shape up	Games and puzzles	Measures, shape and space	Shape and space	4 6	7 8	163 164
109	Right directions	Practice exercise	Measures, shape and space	Shape and space	4 6	9 10	164 165
110	Angle challenge	Investigation	Measures, shape and space	Shape and space	4 6	11 12	165 166
111	Multiple sort	Investigation	Numbers and the number system	Number sequences	8	1 3	177 179
112	The rule of three	Investigation	Numbers and the number system	Number properties	8	1 3	178 179
113	Is it a multiple?	Timed practice exercise	Numbers and the number system	Number properties	8	3	179
114	Tables square	Timed practice exercise	Calculations	Mental calculations × and ÷	9 10	1 2	183 184
115	Double it!	Maths to share	Calculations	Mental calculations × and ÷	9 10	4	184 185
116	Times it!	Timed practice exercise	Calculations	Paper and pencil × and ÷	9 10	5	185 186
117	Divide it!	Practice exercise	Calculations	Paper and pencil × and ÷	9 10	6 8	186 187
118	Remainders	Practice exercise	Calculations	Paper and pencil × and ÷	9 10	6 8	186 187
119	Party shopping game	Games and puzzles	Calculations	Understanding × and ÷	9 10	9	187 188
120	Fraction decimal match	Practice exercise	Numbers and the number system	Fractions	11	1 2	190 191
121	Highest/lowest	Games and puzzles	Numbers and the number system	Decimals	11	1 2	190 191
122	Decimal hunt	Investigation	Numbers and the number system	Decimals	11	3 4	192 193
123	Money adds	Practice exercise	Calculations	Paper and pencil + and −	12	1 3	197 199
124	Shopping check!	Games and puzzles	Calculations	Paper and pencil + and −	12	1 3	197 199
125	TV times	Maths to share	Measures, shape and space	Measures	12	4	199 200
126	Carroll sort	Investigation	Handling data	Organising data	13	1 2	202 203
127	Which circle?	Practice exercise	Handling data	Organising data	13	3	203 204
128	Houses and homes	Maths to share	Handling data	Organising data	13	4 5	204 205

REFERENCE GRID

100 MATHS HOMEWORK ACTIVITIES

100 Maths Homework Activities is a series of teachers' resource books for Years 1–6. Each book is year specific and provides a core of homework activities for mathematics within the guidelines for the National Numeracy Strategy in England. The content of these activities is also appropriate for and adaptable to the requirements of Primary 1–7 in Scottish schools.

Each book offers three terms of homework activities, matched to the termly planning in the National Numeracy Strategy *Framework for Teaching Mathematics* for that year. In schools in England which decide not to adopt the National Numeracy Strategy or for schools elsewhere in the UK, the objectives, approaches and lesson contexts will still be familiar and valuable. However, you will need to choose from the activities to match your own requirements and planning.

These books are intended as a support for the teacher, school mathematics leader or trainee teacher in providing suitable homework activities. The series can be used alongside its companion series, *100 Maths Lessons and more*, or with any maths scheme of work, as the basis of planning homework activities throughout the school, in line with the school's homework policy. The resources can be used by teachers with single-aged classes, mixed-age classes, single- and mixed-ability groups, and for team planning of homework across a year or stage. You may also find the activities valuable for extension work in class or as additional assessment activities.

Using the books

The activities in this book for Year 4/Primary 4–5 classes have been planned to offer a range of mathematics activities for a child to carry out at home. Many of these are designed for sharing with a helper, who can be a parent, another adult member of the family, an older sibling or a neighbour. The activities include a variety of mental arithmetic games, puzzles and practical problem-solving investigations. There are also practice exercises, some 'against the clock', to keep arithmetic skills sharp. The activities have been chosen to ensure that each strand and topic of the National Numeracy Strategy *Framework for Teaching Mathematics* is included, and that the children have opportunities to develop their mental strategies, use paper and pencil methods appropriately, and use and apply their mathematics in solving problems.

Each of the 100 homework activities in this book includes a photocopiable page to copy and send home. Each sheet provides instructions for the child with a brief explanation of the activity for a helper stating simply and clearly its purpose and suggesting support and/or a further challenge to offer the child. The maths strand and topic addressed by each activity and the type of homework being offered are indicated on each page. The types are shown by the following symbols:

maths to share	games and puzzles	practice exercise	investigation	timed practice exercise

There is a supporting teacher's note for each activity. These notes include:

- **Learning outcome:** the specific learning objective of the homework (taken from the National Numeracy Strategy *Framework for Teaching Mathematics*);
- **Lesson context:** a brief description of the classroom experience recommended for the children prior to undertaking the homework activity;
- **Setting the homework:** advice on how to explain the work to the children and set it in context before it is taken home;
- **Back at school:** suggestions for how to respond to the returned homework, such as discussion with the children or specific advice on marking, as well as answers, where relevant.

Supporting your helpers

Extensive research by the IMPACT Project (based at University of North London) has demonstrated how important parental involvement is to children's success in maths. A homework diary photocopiable sheet is provided on page 8 that can be sent home with the homework. This sheet has room for records of four pieces of homework and can be kept in a file or multiple copies stapled together to make a longer-term homework record. For each activity, there is space to record of the name of the activity and the date when it was sent home, and spaces for a brief comment from the helper, the child and the teacher on their responses to the work. The homework diary is intended to encourage home–school links, so that parents and carers know what is being taught in school and can make informed comments about their child's progress.

Name _____

Name of activity & date sent home	Helper's comments	Child's comments		Teacher's comments
		Did you like this? Colour a face.	**How much did you learn?** Colour a face.	
WHERE IS IT COLDEST?	NITA ENJOYED SEARCHING FOR THE INFORMATION FOR THIS ACTIVITY. SHE ESPECIALLY ENJOYED TRYING TO FIND VERY HIGH AND VERY LOW TEMPERATURES.	a lot 🙂 a little 😐 not much 🙁	a lot 🙂 a little 😐 not much 🙁	You have found some interesting temperatures and have shown that you can order them correctly. Well done.
ABACUS CHART	ANTHONY WAS ABLE TO COMPLETE THE CHARTS ON HIS OWN AND THEN I DISCUSSED THE ANSWERS WITH HIM. HE FOUND IT EASIER WHEN HE SAID THE NUMBERS OUT LOUD AS HE WROTE THEM.	a lot 🙂 a little 😐 not much 🙁	a lot 🙂 a little 😐 not much 🙁	Well done. Anthony has a good understanding of place value using four digits. He is ready to move on to five digit numbers now. Please try some out with him at home when you can fit it in. Use both figures and words.

100 MATHS

Using the activities with *100 Maths Lessons series*

The organization of the homework activities in this book matches the planning grids within *100 Maths Lessons: Year 4* (also written by Sonia Tibbatts and John Davis and published by Scholastic), so that there is homework matching the learning objectives covered in each unit of work in each term. Grids are provided on pages 2–5 giving details of which lessons (or series of lessons) in *100 Maths Lessons: Year 4* have associated homework activities in this book along with the relevant page numbers, to help teachers using *100 Maths Lessons: Year 4* for planning.

About this book: Year 4/Primary 4–5

These activities are aimed at nine-year-olds. They are distributed between five strands for the teaching of mathematics to this age group and contain the five types of homework recommended by the NNS. These are 'Practice exercises', some as 'Timed practice exercises' against the clock, 'Maths to share', 'Games and puzzles' and 'Investigations'.

The type of calculations involved in the two kinds of practice exercises form the majority of tasks, as this is an important element of maths work for children in Year 4/Primary 4–5. Working against the clock, though, is not over-emphasised at this stage and will follow later once the children have had chance to reinforce and consolidate their skills and understanding.

Among other key objectives for this year is the need for children to be able to work with larger numbers, up to at least five digits and have the ability to round off to the nearest ten or hundred. More formal methods of recording, especially in addition and subtraction, where numbers are transferred from the horizontal to the vertical and set in columns, are introduced. These include the compensation method in addition, decomposition in subtraction and the grid method and partitioning in multiplication. The homework activities also feature other important number work issues including equivalence in fractions and the introduction of decimals.

'Maths to Share' tasks are intended to encourage the child and a helper to work together on activities at home. 'Games and Puzzles' provide opportunities for children to use and apply their mathematical skills and be systematic in their approach. The 'Investigations' are more open-ended in nature and should form the basis of discussion about the use of strategies and different possible solutions, not only in the home environment, but also when the child returns to school. Because of their format, these tasks might be better set over a weekend or a short holiday.

Name _____

Name of activity & date sent home	Helper's comments	Child's comments		Teacher's comments
		Did you like this? Colour a face. ☺ a lot 😐 a little ☹ not much	**How much did you learn?** Colour a face. ☺ a lot 😐 a little ☹ not much	
		☺ a lot 😐 a little ☹ not much	☺ a lot 😐 a little ☹ not much	
		☺ a lot 😐 a little ☹ not much	☺ a lot 😐 a little ☹ not much	
		☺ a lot 😐 a little ☹ not much	☺ a lot 😐 a little ☹ not much	

Teachers' notes

TERM 1

p29 ABACUS CHARTS
PRACTICE EXERCISE

Learning outcome
* Read and write whole numbers to at least 10 000 in figures and words, and know what each digit represents.

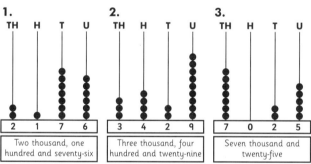

Lesson context
Write three-, then four-digit numbers for the children to read out. Write the numbers in words alongside. Now write the numbers as words and ask the children to provide the digits. Ask them to use digit cards to make numbers to write and draw on to abacus charts, completing the beads, digits and words.

Setting the homework
Demonstrate further examples. Remind the children to complete this task independently, but to discuss their answers with a helper.

Back at school
Swap sheets to check answers together. Check the zero has been used and correctly positioned where needed. Encourage the children to say the numbers aloud.

Answers:

1.

TH	H	T	U
2	1	7	6

Two thousand, one hundred and seventy-six

2.

TH	H	T	U
3	4	2	9

Three thousand, four hundred and twenty-nine

3.

TH	H	T	U
7	0	2	5

Seven thousand and twenty-five

4. 1 252: one thousand, two hundred and fifty-two; **5.** 6 310 six thousand, three hundred and ten; **6.** 4 702 four thousand, seven hundred and two.

p30 ADD OR SUBTRACT
MATHS TO SHARE

Learning outcome
* Add/subtract 100 or 1 000 to/from any integer.

Lesson context
Revise adding and subtracting ones and tens from numbers up to 10 000 according to the children's abilities. Then ask what 100 (then 1 000) more or less than the given numbers will be. Count on/back in steps of 1 000. Finally round integers less than 1000 to the nearest ten or hundred.

Setting the homework
Check the children are able to add/subtract 100 and 1 000 by asking questions like: *What is 100 more than 3 561? What is 100 more than 2 978?* and then subtractions with a zero in the hundreds column: *What is a hundred less than 5 015?*

Back at school
Check the answers with the children, particularly where adding/subtracting 100 changes the value of other digits in the number. Take the opportunity to ask the children to round their answers to the nearest ten/hundred.

p31 JUMPING KANGAROO
GAMES AND PUZZLES

Learning outcome
* Count on or back in hundreds and thousands.

Lesson context
Tell the children to start at zero and count on in hundreds. When you clap, the children count back from that number. Reverse the direction of the count on each clap. Repeat with thousands. Ask the children to complete number lines, counting in hundreds or thousands from a given number.

Setting the homework
Discuss how to find the pattern of a number sequence. Stress that once a pattern has been found, missing numbers should be predicted and then checked to see if they are correct.

Back at school
Once the answers have been checked, ask the children to provide their own number patterns for the rest of the class to find. Ask volunteers to lead the clapping game.

Answers:
1. 577, 477, 377; **2.** 7 121, 8 121, 9 121; **3.** 5 520, 6 520, 7 520; **4.** 6 525, 6 625, 6 725; **5.** 5 117, 4 117, 3 117; **6.** 6 046, 6 146, 6 246; **7.** rule is + 100 + 1 000, next three numbers are 2 624, 3 624 and 3 724; **8.** rule is − 1 000 − 100, next three numbers are 5 345, 5 245, 4 245.

p32 NUMBER SPLITS
PRACTICE EXERCISE

Learning outcome
* Partition numbers into hundreds, tens and ones when adding.

Lesson context
Work through examples of addition on the board using the partition method, HTU + TU then HTU + HTU. Show how the partition method can be written vertically as well as horizontally as an introduction to more formal methods of addition later.

Setting the homework
Work through examples of addition using this method. Remind the children that once the number has been partitioned it then has to be recombined to find the total. Revise methods of setting down, especially converting questions from the horizontal to the vertical format.

Back at school
Invite children to use the board to show others how they used the partition method. Different children could do separate parts of the process. Ask them to talk through the process as they go. Discuss how this method could be used for subtraction.

Answers:
1. 574, **2.** 908, **3.** 430, **4.** 746, **5.** 564, **6.** 785.

p33 IN A WORD — GAMES AND PUZZLES

Learning outcome
- Consolidate knowing by heart: addition and subtraction facts for all numbers to 20.

Lesson context
Children make up addition and subtraction questions with answers up to 20. Explain the meaning of key words used to indicate the two operations.

Setting the homework
Revise the key words used in the activity with the children. List them on the board. Tell the children that once they know what operation is involved, they should try to work out the answers quickly mentally and then check them.

Back at school
Check the children have found the right operation for each question and have calculated the answers correctly. A prominent classroom display of tricky key words might help to jog their memories in the future.
Answers: 1. 15; **2.** 8; **3.** 9; **4.** 19; **5.** 20; **6.** 15; **7.** 9; **8.** 6; **9.** 17; **10.** 13.

p34 ALL CHANGE — MATHS TO SHARE

Learning outcome
- Use addition and subtraction to solve word problems involving numbers in 'real life' and money using one or more steps.

Lesson context
Work through prepared word problems showing children how to pick out the number processes from the words in order to solve the problem. Show examples of one step and two step processes.

Setting the homework
Make sure the children know the coins we currently use in our money system: 1p, 2p, 5p, 10p, 20p, 50p, £1.00 and £2.00. Practise some examples of making up amounts of money in the least number of coins.

Back at school
Check the work and discuss possible alternative answers. Compare the strategies used. Provide them with three/four coins and ask them to make up a money story involving them.
Answers: Section A: 27p = 20p, 5p, 2p; 49p = 20p, 20p, 5p, 2p, 2p; 61p = 50p, 10p, 1p; 84p = 50p, 20p, 10p, 2p, 2p; £1.17 = £1, 10p, 5p, 2p; £3.56 = £2, £1, 50p, 5p, 1p.
Section B: 1. 10p, **2.** 41p, **3.** 25p, **4.** 52p **5.** £1.45.

p35 ODD AND EVEN — GAMES AND PUZZLES

Learning outcome
- Recognise the outcome of sums or differences of pairs of odd/even numbers.

Lesson context
Add and subtract odd and even numbers to establish general rules. Answers are checked using the inverse operation. Work through some examples on the board. Start with addition of small two-digit numbers. Then check subtraction.

Setting the homework
Ensure the children are conversant with odd and even. Talk about house numbering where there are dwellings on both sides of the street. Stress that to prove number rules they may need to test out more examples than there are on the sheet.

Back at school
Make sure the children have the correct answers and shaded in the right squares to make the capital letters O and E. Check through the rules for adding and subtracting odd and even numbers. *Do the rules form a pattern? Did anyone find any exceptions? Test them out.*

p36 DIFFERENT ORDER — INVESTIGATION

Learning outcome:
- Check the sum of several numbers by adding in reverse order.

Lesson context
Children solve some word problems involving 'real life' situations by adding numbers together. The order of the numbers is then changed. Does changing the order of the numbers change the result?

Setting the homework
Use real life examples of adding two-digit numbers from inside the classroom. Check the results of totals by changing the order of the numbers and adding them again. How many different ways are there? Ask them to predict what their results will show.

Back at school
Invite children to give examples of real life additions that they have done at home. What results did they get when the order of the numbers was changed around? Is there a definite rule? Did their prediction turn out to be true?

p37 MEASURE UP — PRACTICE EXERCISE

Learning outcome:
- Use, read and write standard metric units of length including their abbreviations.

Lesson context
Revise the metric units used to measure length. Write the words and their abbreviations on the board. Give each child a 30cm ruler and discuss how each centimetre is divided into 10mm with 5mm as the half-way point. Show how to read off to the nearest half centimetre.

Setting the homework
If any child does not have a 30cm centimetre ruler it may be possible for them to take their school ruler home. Revise reading measurements to the nearest half centimetre and possibly how to write measurements in at least two different ways.

Back at school
Check measurements to the nearest half centimetre. Check answers to the nearest millimetre for those who have measured more accurately. Discuss how the lengths have been written in a number of different ways. Go for other alternatives where possible. **Answers:** 10.2cm; 8.6cm; 3.0cm; 2.4cm; 3.2cm; 7.7cm; 1.0cm; 3.0cm.

p38 LONG DISTANCE — INVESTIGATION

Learning outcome
- Suggest suitable units and measuring equipment to estimate and measure length.

Lesson context
Children estimate measuring tasks in the classroom, then take measurements using a range of equipment. They record their results on a chart, giving answers to the nearest centimetre. Examine the types of scales and units used to record the results.

Setting the homework
Ensure the children have access to a metre tape measure. Talk about estimating and encourage them to be honest even if some attempts are inaccurate. Work through examples of rounding up or down to the nearest centimetre.

Back at school
Talk about the range of objects the children chose to measure and key words like 'length', 'width' and 'height'. Discuss how close their estimates were to the actual measurements. What problems were encountered using the scale marked on the metre tape measure?

p39 COVER UP

Learning outcome
- Calculate the perimeter and area of rectangles and other simple shapes using counting methods and standard units.

Lesson context
Discuss the differences between area and perimeter. Demonstrate how to calculate the area of simple shapes by counting squares. Show how to record this using the units square centimetres (cm²). Look at quicker ways of calculating both area and perimeter by calculation.

Setting the homework
Remind children about the differences between area and perimeter. Draw several rectangles on the board. Demonstrate how to count squares to find area and to calculate the perimeter by adding lengths of sides together.

Back at school
Check through the answers. Concentrate on the strategy the children used to find area. Lead towards the area of a square or rectangle being equal to length × breadth. Also consider formula for finding perimeter more quickly.

p40 ALL SHAPES AND SIZES

Learning outcome
- Describe and visualise 2-D shapes; classify them according to their properties.

Lesson context
Revise the main 2-D shapes. Discuss the properties of these shapes and key vocabulary. Discuss the meaning of 'regular' and 'irregular' in the context of these shapes. Introduce the word 'polygon'. Introduce the heptagon and revise the pentagon, hexagon and octagon.

Setting the homework
Go through the list of shapes and tell the children they have to find examples of these shapes inside and outside their homes. Encourage accurate drawing and labelling and tell the children they should use a ruler where appropriate.

Back at school
Check on the range of 2-D shapes collected. Talk about the properties of each, especially the number of sides, angles and corners. Check that the children understand the difference between regular and irregular shapes and that they are able to give a definition of the word 'polygon'.

p41 IN THE NET

Learning outcome
- Make shapes and discuss properties.

Lesson context
Return to the cube. Display the large version of the net. Ask the children to draw it, cut it out and decorate it ready for fixing together. Try using pieces of artstraws to make the cube skeleton.

Setting the homework
Revise the work done on making the cube from a net. Explain the children are going to do a similar task with the tetrahedron or triangular-based pyramid. They will be given one net to make up and then asked to find other nets for the same shape. Provide the children with extra sheets of triangular grid paper to try out their nets.

Back at school
Make a display of the nets and/or shapes that the children have brought into school. How many different successful nets did they find? Look at the properties of the shape and compare and contrast it with the Egyptian pyramid.

p42 DOUBLE TROUBLE

Learning outcome
- Derive quickly doubles of all numbers to 50 and the corresponding halves.

Lesson context
Stress that halving is the inverse of doubling, then give quick-fire questions on doubles of numbers up to 50. Repeat for halving, starting with even numbers then include odd numbers. Create number chains doubling up from a starting number. Repeat for halving, stopping when a fraction is reached.

Setting the homework
Remind the children that doubling and halving are inverse operations. Revise partition strategy for doubling and halving, e.g., double 27 is double 20, then double 7, 40 + 14 = 54 and half of 48 is half of 40 and half of 8, 20 + 4 = 24.

Back at school
Once the answers have been checked, see how many of the numbers between 0 and 50 the children can double or halve quickly in their heads. Discuss the strategies they use. Look at alternatives to the partition method mentioned previously.

p43 TIMES THREE

Learning outcome
- **Know by heart multiplication facts for the 3 times table; derive quickly division facts for the corresponding table.**

Lesson context
Chant through the 3 times table from 3 to 30 and then back again. Ask random questions based on the table. Leave gaps in the 3 times table when writing it on the board and ask the children to fill them in. Ask questions that involve division facts. Reinforce knowledge of tables using ×3 function machines.

Setting the homework
Quickly revise the 3 times table in various ways involving both multiplication and division. Ask questions such as: *What number on the 3 times table comes after 24? What number comes before 15? What is 8 times 3? Divide 30 by 3. How many threes are there in 24?*

Back in school
Check to see what the children have discovered. Invite some children to give examples of the multiplication and division statements they have written. How far can the children extend the 3 times table by counting in threes beyond 30? Discuss the pattern formed by adding the digits making up the multiples of 3.

p44 THROUGH THE MAZE

Learning outcome
- **Know by heart multiplication facts for the 3 and 4 times tables.**

Lesson context
Play a tables bingo game using questions from the 3 and 4 times tables. Include questions that reinforce the fact that division is the inverse of multiplication.

Setting the homework
Chant through the 3 times table from 0 to 30 and then in reverse. Test the children's knowledge of the 3 times table by asking random questions. Repeat the process for the 4 times table.

Back at school
Put up large versions of the number mazes on the board or as a wall-chart. Plot the trail through the mazes with the children, asking them for suggested routes. Provide them with blank squared paper and challenge them to devise their own numbers mazes based on the 3 and 4 times tables.

p45 MAGIC SQUARES — INVESTIGATION

Learning outcome
- Make and investigate a general statement about familiar numbers.

Lesson context
Explain what magic squares are. Add up each line in the example to show it totals 15. On squared paper, ask the children to rearrange the same nine digits in the square to make it remain a magic square. Each digit can only be used once. How many different solutions can they find? Ask them to make a general statement about magic squares.

Setting the homework
Work through examples using single-digit numbers, showing that if two numbers in the line are already known, the third number can be calculated by adding them and then subtracting from the known total.

Back at school
Work through the magic squares, with the children providing the missing numbers. Discuss the strategies they used using fractions and decimals. Ask them to provide magic squares they have made up. Show them that the square will remain a magic square providing they follow the same rule with every number, such as 'add 3', or 'double every number'.

p46 IN THE FAMILY — PRACTICE EXERCISE

Learning outcome
- Extend understanding of the operation of multiplication and its relationship to adding and dividing.

Lesson context
Establish that multiplication is a quick way of adding. Work examples where multiplication can be used to check answers. Show examples where any number × 1 leaves the number unchanged and any number times zero is always zero.

Setting the homework
Reinforce the fact that once the children have been given one multiplication fact, a second multiplication fact and two division facts can be derived from it. Emphasise the phrase 'inverse operation' and stress how useful this is when checking answers to multiplications and divisions. Give facts orally and ask the children to say the related facts back to you.

Back at school
Check that the children have pasted the strips into the correct multiplication and division families. Which ones did they find most difficult to do? Extension: could they apply the same strategies when the numbers become larger?
Answers: 6 × 2 = 12, 12 ÷ 2 = 6, 12÷ 6 = 2, 2 × 6 = 12;
9 × 3 = 27, 27 ÷ 3 = 9, 27 ÷ 9 = 3, 3 × 9 = 27;
9 × 5 = 45, 45 ÷ 5 = 9, 45 ÷ 9 = 5, 5 × 9 =45;
5 × 10 = 50, 50 ÷ 10 = 5, 50 ÷ 5 = 10, 10 × 5 = 50;
7 × 4 = 28, 28 ÷ 7 = 4, 28 ÷ 4 = 7, 4 × 7 = 28.

p47 NUMBER TRIOS — MATHS TO SHARE

Learning outcome
- Extend understanding of the operation of division and its relationship to subtraction and multiplication.

Lesson context
Tell the children that dividing can be carried out using repeated subtraction. Show examples. Also work examples where division is shown to be the inverse of multiplication. Give examples to show that any number divided by 1 leaves the number unchanged.

Setting the homework
Reinforce the fact that once the children have been given one division fact, a second division fact and two multiplication facts can immediately be derived from it. Again stress the importance of using the 'inverse operation' when checking answers. Give some examples of the type of question given on the homework sheet.

Back at school
Check the answers, ensuring that each number trio has produced two division and two multiplication facts based on the same numbers. Revise some key words connected with multiplication (e.g., times and product) and those used in place of 'division' (e.g., share and group). **Answers:** 5 × 2 = 10, 2 × 5 = 10, 10 ÷ 5 = 2, 10 ÷ 2 = 5; 9 × 10 = 90, 10 × 9 = 90, 90 ÷ 10 = 9, 90 ÷ 9 = 10 and so on.

p48 DOUBLE CROSS — GAMES AND PUZZLES

Learning outcome
- Use doubling and halving, starting from known facts.

Lesson context
One way to multiply by 4 is to double and double again. Similarly one way to multiply a number by 5 is to multiply by 10 and halve the answer. One way to multiply by 20 is to multiply by 10 and then double the answer. Let the children use digit cards to produce their own single-digit and two-digit numbers to work examples of these three methods.

Setting the homework
These methods will enable the children to work out answers quickly in their heads. Revise the rules. Work through some further examples and get the children to answer orally.

Back at school
Check the children have used the methods taught. Monitor their response times to see how quickly they can work out the answers. Can the children extend what they have learned to help speed up multiplying by other numbers?
Answers: 1a. 52; **1b.** 76; **1c.** 148; **1d.** 104; **1e.** 168. **2a.** 240; **2b.** 640; **2c.** 480; **2d.** 940; **2e.** 1100. **3a.** 130; **3b.** 285; **3c.** 390; **3d.** 215; **3e.** 180.

Learning outcome
• Use known number facts and place value to multiply and divide by 10.

Lesson context
A quick way to multiply whole numbers by 10 is to move the digits one place to the left, leaving the units column with a zero. Write up statements with missing numbers for the children to solve. A quick way to divide a number that ends in 0 by 10 is to move the digits one place to the right, which means there is no 0 in the answer. Again provide missing number statements for them to solve.

Setting the homework
Work through examples of the two rules, asking the children to give their answers orally. Stress that the multiplying by 10 rule only applies to whole numbers and that the dividing by 10 rule should only be used if the number being divided ends in zero.

Back in school
Time how quickly the children can respond. Ask children to make up statements with missing numbers for the others to solve. Can they suggest quick ways of multiplying and dividing by 100?

Answers:
× 10: 190, 470, 940, 1250, 3520, 5940; ÷ 10: 7, 15, 21, 39, 652, 874.
Missing numbers: **1.** 23, **2.** 10, **3.** 654, **4.** 7150, **5.** 2170, **6.** 10, **7.** 2300, **8.** 10, **9.** 943, **10.** 759.

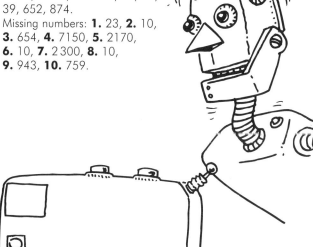

Learning outcome
• Approximate first. Use informal pencil and paper methods to support, record and explain multiplication.

Lesson context
Remind the children of the grid method introduced in Year 3. Use 2, 3, 4, 5 and 10 times tables. Ask them to approximate first and then show how to multiply using the grid method, partitioning the two-digit number to multiply the tens and units separately and then adding together the results to give the answer. Work through other examples. Use tables squares to check calculations.

Setting the homework
Make sure the children understand the process. Stress the importance of obtaining an approximate answer first by rounding off. Give access to a tables square, if necessary. Ask the children to think about another method they might use to check their answers.

Back at school
Discuss how useful they found the method. What parts of it did they find most difficult? How close was their approximate answer? Did it provide a useful guide? What other methods did they use to check their answers. **Answers: 1.** 135, **2.** 237, **3.** 118, **4.** 128, **5.** 380, **6.** 504, **7.** 135, **8.** 2985.

Learning outcome
• **Find remainders after division.**

Lesson context
Remind the children that remainders are always written as a whole number. Work through some examples. Show the children how these can be written in a different way. Explain that it is sometimes necessary to round down or round up to get a sensible answer. Children work through questions using the 'multiply and add the remainder' method to check their answers.

Setting the homework
Work through examples of divisions that produce remainders. Show the children how to check their answers by multiplying and then adding on the remainder. Make sure those who need tables squares have copies to take home.

Back at school
Ask one child to show how they worked out the answers. Ask another child to check the answer using the suggested method. Discuss 'real life' situations where remainders might occur when dividing numbers.

Answers: 1. 3 r 3, 2. 2 r 4,
3. 7 r 2, **4.** 3 r 1, **5.** 3 r 6, **6.** 3 r 5, **7.** 2 r 5, **8.** 4 r 7, **9.** 6 r 8, **10.** 4 r 5, **11.** 8 r 2, **12.** 6 r 9. Message reads 'very good work'.

Learning outcome
• Use all four operations to solve word problems involving numbers in 'real life' including money.

Lesson context
Give children some single process word problems, then move on to two-step questions. Also use examples where it is necessary to round up/down following a division question with a remainder. Children then complete a money/shopping activity.

Setting the homework
Run through more examples where children have to pick out the processes, numbers and operations from a word problem. Look at one step and two step problems. Discuss key words and phrases often found in money problems.

Back at school
Discuss the children's answers. Were they able to use an inverse operation to check their answers? Ask them to make up their own money 'real life' problems. Answers and possible combinations of coins can be easily checked.

p53 SPOT THE FRACTION
PRACTICE EXERCISE

Learning outcome
• Use fraction notation.

Lesson context
Demonstrate that 'half' means 'one equal part out of two' and that 'one quarter' means 'one equal part out of four'. Repeat the process for other fractions. Children make their own fraction strips using squared paper.

Setting the homework
Establish that the children understand that a fraction is part of a whole one. Show the children some examples of drawing squares and rectangles using squared paper and dividing the shapes up into different fractions.

Back at school
Children show the diagrams they have drawn to show certain fractions. How many different variations are there? Reinforce that the coloured part of the shape and the part left blank go together to make one whole one. **Answers:** 1–8 children's own shapes. Counters: two-eighths green, one-eighth red, three-eighths yellow, two-eighths not coloured. Balloons: children's own answers.

p54 FAIR SHARES
MATHS TO SHARE

Learning outcome
• Begin to relate fractions to division and find simple fractions of numbers and quantities.

Lesson context
Explain that fractions can be linked to division. Show that finding one half of a number is the same as dividing by 2 and that when a whole cake is divided into four equal parts, each piece is one quarter of the cake. Give examples of finding fractions of certain amounts such as: *What fraction of £1.00 is 25p?*

Setting the homework
Emphasise the link between fractions and division. Stress that we find a half by dividing by 2 and a quarter by dividing by 4. Run through some examples and show how to convert the word problem into a diagram.

Back at school
Check the children's answers. How did the diagrams help? Which questions did they find most difficult to put into diagram form? Talk about fraction questions related to money and metric measures. **Answers: 1.** 5 apples; **2.** 9 biscuits; **3.** 4 pears; **4.** 60p; **5.** 9 sweets; **6.** 7 books carried, 13 books left.

p55 HIDDEN FRACTIONS
GAMES AND PUZZLES

Learning outcome
• Begin to relate fractions to division and find fractions of numbers and quantities.

Lesson context
As previous lesson. Provide a paper-and-pencil task where the children are finding halves, fifths, quarters and tenths of numbers, money and/or measures.

Setting the homework
Explain that the first part of the activity is to help them think about the relationship between different fractions. Work through some examples. Show simple number lines running from zero to one, and show how to position fractions.

Back at school
Invite children to explain how they worked out the answers. Discuss how to place fractions in order of size.
Answers: Fraction shapes: half, quarter, three-quarters, two-thirds. Letters in the same place will be A, a half, and H, four-eighths, because they are equivalent fractions. Order of size for fractions: one-eighth, quarter, three-eighths, four-eighths and a half, five-eighths, three-quarters, seven-eighths.

p56 QUICK CHECK
PRACTICE EXERCISE

Learning outcome
• Consolidate understanding of the relationship between + and – and check with the inverse operation.

Lesson context
Use number questions to test the children's subtraction vocabulary. Reinforce the fact that subtracting zero from any number leaves the number unchanged. Children generate their own numbers for subtraction questions which they can then check by adding.

Setting the homework
Revise words and phrases commonly used to indicate a subtraction. Show examples where a subtraction can be checked by adding. Discuss the term 'inverse operation' and explain what it means.

Back at school
Go through the answers to ensure subtractions have been carried out and checked successfully using the vertical format. Try some examples of HTU – TU. **Answers: 1.** 57; **2.** 37; **3.** 36; **4.** 168; **5.** 98; **6.** 159.

p57 NUMBER WORKOUT
GAMES AND PUZZLES

Learning outcome
• Use informal pencil and paper methods to support, record and explain additions/subtractions.

Lesson context
Revise the children's mental calculation strategies and informal written methods of adding and subtracting. Children then work through prepared questions.

Setting the homework
Talk through some examples with missing numbers involving addition and subtraction and ask for suggested strategies for finding. Remind children to use the inverse operation to check their answers.

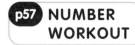

Back at school
How did they solve the puzzle and find the number? Talk about the different strategies used. Ask children to make up their own puzzles with up to two missing numbers for others to solve.
Answers: 1. 45 + 77 = 122; **2.** 96 – 32 = 64; **3.** 84 – 39 = 45; **4.** 125 + 78 = 203; **5.** 467 – 31 = 436; **6.** 279 – 127 = 152; **7.** 693 + 254 = 947; **8.** 536 – 273 = 263.

p58 JUST IN TIME
MATHS TO SHARE

Learning outcome
• Read the time from an analogue clock to the nearest minute, and from a 12-hour digital clock.

Lesson context
Revise time vocabulary and units of measure. Using a large clock-face, remind the class of what the hour and minute hands represent. Run through times past o'clock and times to o'clock. Discuss a.m. and p.m. times. Show how times can be written in different ways. Explain the difference between an analogue clock-face and a digital clock-face. Let the children work on telling the time tasks.

Setting the homework
Revise information covered in the lesson above to ensure children have understood the main concepts.

Back at school
Invite children out to show the answers to the questions in two different ways where possible. Discuss key vocabulary.
Answers: 1. Check correct positions of hands on the clock face. **2.** 8:40, 6:15, 4:00 and 10:50. **3.** 10 past 8, 25 minutes past 9; 6 o'clock; ten minutes to 1. **4.** 25 minutes past 3, 5 minutes past 9; quarter past 2; 5 minutes to 12.

p59 BUSY DAY   INVESTIGATION

Learning outcome
• Use, read and write the vocabulary related to time.
Lesson context
Revise information covered in previous lesson.
Setting the homework
Discuss the main events during a normal school day. Explain to the children that they are going to make a time log to record what these main events are, at what time they start and approximately how long they last. Talk through how to complete the chart.
Back at school
Compare what the children have recorded on their charts. Discuss the methods and strategies used to work out how long events lasted. Which type of clock did they find easier to use, analogue or digital?

p60 FINDING OUT   PRACTICE EXERCISE

Learning outcome
• Solve a problem by collecting quickly, organising, representing and interpreting data.
Lesson context
Discuss data handling and what each stage of the process entails. Explain that they are going to make a database about themselves. Hand out personal record charts and ask them to record the information. The class take turns to collate their information on the computer database. Once complete, the children take turns to use the database to answer the questions they have devised.
Setting the homework
Run through the homework sheet. Explain how the spare columns can be used to help collate and sort the information. Discuss some possible uses of these columns. Make sure that the children understand the questions. Discuss the use of terminology such as 'all round' and 'half marks'.
Back at school
Talk about the answers and discuss the methods and strategies they used to find solutions. Which answers proved to be most difficult to find? Would it be useful to produce their own class test results in this way?
Answers: 1. Aslam; **2.** Sita; **3.** Sita; **4.** Paula; **5.** 8; **6.** 7; **7.** 4; **8.** 3; **9.** False; **10.** This requires discussion! Ask: *Is Paula's score of* $^{10}/_{20}$ *better or worse than* $^{13}/_{30}$*?*

p61 HAPPY BIRTHDAY   MATHS TO SHARE

Learning outcome
• Solve a problem by collecting quickly, organising, representing and interpreting data using a tally chart.
Lesson context
Show the children on the board how to record items in a count using tallying. Gather information from the class, for example on the pets children keep. The children then work in groups to devise questions to interrogate the tally chart.
Setting the homework
Check that the children understand how to count up the items in the tally chart to find the total number of birthdays for each month of the year. Revise the meaning of the symbols that are used. Discuss some examples.
Back at school
Make sure the children have counted up correctly. Run through the questions. Discuss the possible alternatives given for question 8 where they had to devise their own problem.
Answers: 1. 5; **2.** 2; **3.** April; **4.** May, July, Oct, Dec; **5.** June; **6.** 6; **7.** 13.

p62 GET INTO SHAPE   MATHS TO SHARE

Learning outcome
• Solve a problem by collecting quickly, organising, representing and interpreting data using pictograms.
Lesson context
Check the children know what a pictogram is. Explain that the diagram or symbol used can represent one, or more than one, unit. In pairs, the children decide what each symbol should represent and then make a pictogram. The pictogram will need a title and must show how many items each symbol stands for. They then prepare questions that others can answer using the pictogram.
Setting the homework
Explain that the children will make a pictogram from information shown on the homework sheet. Revise pictograms and the stages required to draw one. *If the numbers involved are large, how many shapes should each symbol represent?*
Back at school
Ask the children to demonstrate how they recorded the different shapes counted. Talk about what they have found out from the pictogram. How did the information help the teacher?

p63 SPORTY TYPES   INVESTIGATION

Learning outcome
• Solve a problem by collecting quickly, organising, representing and interpreting data using pictograms.
Lesson context
Children complete a pictogram of their favourite percussion instruments.
Setting the homework
Tell the children they are going to survey their friends, teachers and family to find out about their favourite sporting games and make a pictogram. They should NOT ask anyone else. Put some sample pictures on the board to illustrate the different sports. Squared paper is provided to help them organise their work.
Back at school
Look at some of the children's work. Ask them to explain the stages of making their pictogram. Ask others to give a summary of their findings. In what ways were their results similar/different from other children's? Discuss the advantages and disadvantages of using a pictogram.

TERM 2

p64 COMPARISON SENTENCES
MATHS TO SHARE

Learning outcome
- **Use symbols correctly, including less than (<), greater than (>), equals (=).**

Lesson context
Explain to the children about the use of these symbols. Together work on a range of practice examples, such as 12 > 5, then 5 + 7 > ? And 10 < ? < 25.

Setting the homework
Discuss the example and let them suggest possible 'comparison sentences' and demonstrate how these can be written using the symbols. Tell the children to write a 'word sentence' about the comparison followed by a 'number sentence'.

Back at school
Ask children to share some of their examples and suggest a comparison sentence using 'greater than'. Make sure that the symbols < and > are being used correctly. Tell them to think of the symbol as a crocodile's mouth, the biggest number is on the widest side of the symbol where the 'mouth' is open and the smaller number is on the 'closed' side of the 'mouth'.

p65 WHERE IS IT COLDEST?
INVESTIGATION

Learning outcome
- Recognise negative numbers.

Lesson context
Count from positive to negative numbers in ones, twos and fives, using the class number line. Practise ordering numbers, including negative numbers, using cards on a washing line. Look at negative numbers in the context of temperature and work on ordering them from hottest to coldest.

Setting the homework
Tell the children that you want them to find the temperatures of some places in the world, especially very cold places. Talk to them about the various sources of information. Ask them to try to find about ten different temperatures and then list them in order.

Back at school
Let children share their information. Ask each child for the coldest place they have found and let them order the data. The children use the information to make a list, putting each of the places in order, or plot the information on a graph.

p66 MAKING SUMS
PRACTICE EXERCISE

Learning outcome
- Consolidate understanding of relationship between + and −. Understand the principals (not the names) of commutative and associative laws as they apply or not to addition and subtraction.

Lesson context
Introduce the fact that 87–46 is not the same as 46–87, while addition sums can be done in any order (associative law). Show the children how they can use this knowledge to 'check' their addition by re-arranging sums and adding in a different order, and also by using the inverse operation.

Setting the homework
The purpose of this activity is to develop understanding of the relationship between addition and subtraction and to help the children recognise the different ways a sum can be written. Go through the example on the sheet.

Back at school
Go through the examples and let children mark their own work. Look at 38 and 38, ask how many different sums for this one.

p67 DOUBLE UP
GAMES AND PUZZLES

Learning outcome
- Derive quickly: doubles of all whole numbers to 50.

Lesson context
Work on 'doubling' numbers up to 50 using partitioning strategies. Then, using the information about doubles, find 'near doubles', for example 120 + 130 as double 120 + 10.

Setting the homework
Tell the children that this game allows them to practise their 'doubling' skills. Remind them about the strategies they can use to work out 'doubles'. Make sure that they have a dice. You may like to differentiate the activity by providing children with different dice, for example a ten-sided dice for more able children to make three-digit numbers.

Back at school
Talk through any difficulties individual children had. Children could check each other's work.

p68 WORK IT OUT!
PRACTICE EXERCISE

Learning outcome
- Use informal paper and pencil methods to support, record or explain additions.

Lesson context
In class, question children about the different strategies they use for addition. Remind them to look for the easiest calculation, especially looking for number bonds, doubles and numbers that are close to multiples of ten.

Setting the homework
Tell the children that the exercise contains a range of different additions and you want them to think about the easiest way to calculate each one. Give some examples and discuss the different methods they can use to calculate the answers. Explain that they need only show their calculations informally as jottings.

Back at school
Go through the questions asking children to explain their working. If children have used different methods, compare them, give further examples and discuss which are the most efficient. **Answers: 1.** 67; **2.** 228; **3.** 500; **4.** 130; **5.** 170; **6.** 113; **7.** 120; **8.** 80; **9.** 136; **10.** 178.

p69 SPENDING YOUR BIRTHDAY MONEY
INVESTIGATION

Learning outcome:
- **Choose and use appropriate number operations and appropriate ways of calculating to solve problems.**

Lesson context
In the lesson, teach children about adding money. Remind them of how money is written, either as decimals of a £ or as pounds and pence.

Setting the homework
This homework activity encourages children to think about money addition and make choices within a budget. They will be comparing prices and looking at different combinations of amounts. You can differentiate this activity by modifying the criteria, for example, you may say children can 'round' the amounts to the nearest pound and then total them.

Back at school
Ask children to share some of their examples. Work out the most popular items and compare costs. Talk about the difficulties of having to keep within the budget and get as close to the target amount as possible.

p70 UNITS FIRST — PRACTICE EXERCISE

Learning outcome:
• **Develop and refine written methods for column addition of two whole numbers less than 1000.**

Lesson context
Go through the extended method of addition of HTUs, adding the units first. It may be useful to use Base Ten apparatus to demonstrate this, as it is important that children clearly understand the place value of each digit. Demonstrate some examples where the tens gives value >9.

Setting the homework
Tell the children how you want them to do this practice exercise, stressing that it is important that they use the method you have shown them. Look at the first example and remind them of the value of each of the digits.

Back at school
Go through any questions that the children found difficult, especially when there is a zero in the answer. If the children are confident with this method you may then decide to introduce the shortened version of addition, adding the units first.
Answers: 1. 869; **2.** 897; **3.** 890; **4.** 899; **5.** 900; **6.** 992; **7.** 800; **8.** 794.

p71 EXCHANGE IT! — PRACTICE EXERCISE

Learning outcome:
• **Develop and refine written methods for column subtraction of two whole numbers less than 1000.**

Lesson context
Demonstrate the decomposition method for column subtraction, as illustrated on the homework sheet. Work through some examples with the children, reinforcing their understanding of place value. Stress, for example, that 265 is two hundreds and sixty and five.

Setting the homework
Explain that this is a practice exercise and that you want the children to show their working in the way illustrated, pointing out that they will not always have to exchange both hundreds and tens.

Back at school.
Go through the worksheet looking at common errors. If children are confident with this extended method, introduce the shortened method and give the children some further examples to do using both methods. **Answers: 1.** 319; **2.** 849; **3.** 190; **4.** 198; **5.** 768; **6.** 95.

p72 WHAT MEASUREMENTS DO WE USE? — INVESTIGATION

Learning outcome
• **Know the relationships between familiar units of length, mass and capacity.**

Lesson context
Talk to the children about different units of measurement and how they are recorded. Ask them which units they know and make sure that they know the terms mass, capacity and length, and which units relate to each. Remind them about equivalence of units.

Setting the homework
Explain that you want them to find items in their own home which are marked with units of measurement. Give some examples such as those on the homework sheet, pointing out that they should be recorded under the appropriate heading.

Back at school
Compare children's lists and discuss the units used. Children could work in groups to make lists under each of the headings, including a list of the 'unusual' items. Most will be 'weekly shopping' items. Talk about the way that different quantities are packaged.

p73 HOW LONG IS IT? — GAMES AND PUZZLES

Learning outcome
• Know the equivalents to halves, quarters and tenths of 1km, 1m, 1kg and 1l in m, cm, g and ml. Convert up to 1000cm to metres, and visa versa.

Lesson context
The children should be familiar with the common units of measurement. Remind them that 10mm = 1cm and that 100cm = 1m. Talk about fractions of a metre. Let the children estimate some measurements. Encourage them to estimate in fractions of a metre, giving the equivalent number of centimetres. They should then measure accurately to check their estimates.

Setting the homework
Tell the children that you want them to measure the length of some items around the home, after estimating as they did in class. Suggest they work with someone at home, taking it in turns to estimate a length and then checking each other's measurements.

Back at school
Encourage children to talk about their work and about the items they have measured. Let them play a game, challenging each other to estimate the lengths of items around the classroom.

p74 MEASURING UP! — PRACTICE EXERCISE

Learning outcome
• Use all four operations to solve problems involving capacity.

Lesson context
Remind the children of the different units of measurement. Explain the importance of working with the same units when doing calculations, for example when adding 1.5m to 65cm, add either m or cm. Give children some examples of calculations using units of measurement, stressing the correct use of the decimal point.

Setting the homework
Tell the children that these are 'real life' problems. Ask them to think about the correct operation to use and remind them to pick out the relevant pieces of information from each question.

Back at school
Go through the questions, letting children explain their calculations. Make sure that children are recording their answers correctly and including the units.

Answers: 1. 7m20cm; **2.** 20; **3.** Two bottles; **4.** 46kg; **5.** 675g.

p75 DRAW AND COUNT

GAMES AND PUZZLES

Learning outcome
• Measure and calculate the perimeter and area of rectangles and other simple shapes, using counting methods and standard units (cm, cm²).

Lesson context
Let the children draw some simple shapes on centimetre squared paper and count the number of squares to find the area. Start with rectangles and then introduce shapes which include parts of square centimetres in their area. Explain that the units used are cm². Tell the children that where there are incomplete squares, they must decide whether the part square is more or less than a half, and count it accordingly.

Setting the homework
Children may need some additional squared paper. Tell them to find items that will fit onto the page and draw around them.

Back at school
Let the children show some of their examples. Remind them that if they have used part squares and rounded up or down the area will not be exact, but will be a good approximation.

p76 MAKE A PICTURE

GAMES AND PUZZLES

Learning outcome
• Describe and find the position of a point on a grid of squares where the lines are numbered.

Lesson context
Demonstrate how the axes are numbered on a prepared grid, explaining that this enables us to identify any point using the co-ordinates. Explain how the co-ordinate is arrived at, pointing out that the horizontal co-ordinate is given first. Let the children identify points on the grid and give the co-ordinates of points that you identify.

Setting the homework
Explain that you want them to design a simple picture and to give the co-ordinates of each of the points on the picture, so that someone else can draw it.

Back at school
Select children to give the co-ordinates of their picture for the class to draw. Children should then compare the resulting pictures to check that they are the same.

p77 WHICH DIRECTION?

PRACTICE EXERCISE

Learning outcome

• Recognise and use the eight compass directions.

Lesson context
In class show the children how the eight compass directions are derived. Use a simple map to demonstrate and ask the children to identify directions and describe positions using compass directions.

Setting the homework
Tell the children to look at the plan they have been given. Ask questions such as: *Which is further north Page Park or Uptown School? Whose house is SW of the school?* Explain that the questions they have to answer are similar to these.

Back at school
Go through the answers. Children could then design their own plan and a set of similar questions.

p78 SHAPE COLLECTION

MATHS TO SHARE

Learning outcome
• Make and investigate a general statement about familiar shapes by finding examples that satisfy it.

Lesson context
Talk to children about the properties of different 2-D and 3-D shapes. Let them describe shapes to each other by making general statements about them. For example point out that rectangles can vary in size and shape, but they will all satisfy the general statements.

Setting the homework
Explain to the children that you want them to find objects that are examples of different shapes. Tell them to try to find at least two examples of each shape that are as different as possible, but still match the general statements.

Back at school
Talk about the examples that children have found, reinforcing the concept that the shapes must satisfy general statements. This could be followed up with a 'Guess my shape game', where the teacher gives a shape definition for the children to identify.

p79 FIND THE RULE

MATHS TO SHARE

Learning outcome
• Recognise and extend number sequences formed by counting from any number in steps of constant size, extending beyond zero when counting back.

Lesson context
Ask children to count in steps of 2, 4, 5, 10, both forwards and backwards. Demonstrate some simple number sequences on the board and ask the children to work out 'steps'. Explain that they can work out 'the rule ' in this way. Give some examples of other sequences involving doubling and halving. The children should then play the 'Find the rule' game.

Setting the homework
Tell the children that you want them to play 'Find the rule' with someone at home. Encourage them to try to think of interesting sequences. Discuss the examples on the worksheet.

Back at school
Ask the children to give some examples of the sequences they used – see if other children can spot the rule!

Answer: times 2 or double, subtract 3, halve.

p80 ODDS AND EVENS

GAMES AND PUZZLES

Learning outcome
• Recognise odd and even numbers up to 1000, and some of their properties, including the outcome of sums or differences of pairs of odd/even numbers.

Lesson context
Talk to the children about odd and even numbers, explaining that the critical digit is the last one. Even numbers all end in 0, 2, 4, 6 or 8 and odd numbers end in 1, 3, 5, 7 or 9. Let children investigate the outcomes of adding odd and even numbers.

Setting the homework
Tell the children that the homework activity is a game to help them recognise odd and even numbers quickly. Suggest that they play with someone at home.

Back at school
Remind children how to recognise odd and even numbers, demonstrate with some large numbers >100, >1000. Give the children some addition examples and ask them to tell you whether the answer will be odd or even.

p81 NUMBER PUZZLES

Learning outcome
• Solve mathematical problems or puzzles, recognise and explain patterns and relationships, generalise and predict.

Lesson context
Introduce a variety of number investigations asking children to explain their reasoning. Encourage children to think about how they work problems out.

Setting the homework
Talk about how codes are used. Explain that the code shown on the worksheet is a very simple one. Get them to work out their own name in 'code'. Give them some simple sums 'in code' such as E times E (5 × 5) giving the answer Y (25).

Back at school
Discuss any difficulties that children had with the problem. Let the children share their 'secret messages' for the others to solve. You may like to develop this further by letting the children decide their own codes and send messages.
Answer: 2, 18, 9, 12, 12, 9, 1, 14, 20 – BRILLIANT.

p82 TABLES BINGO

Learning outcome
• **Know by heart multiplication facts for 2, 3, 4, 5 and 10 times tables.**

Lesson context
Let the children colour different multiples on a 100 square. Point out the patterns of the different times tables. Children work in pairs or small groups to practise instant recall of tables, challenging each other with different tables and using multiplication squares to check their answers.

Setting the homework
Show the children the worksheet and talk to them about choosing numbers to complete a 'bingo card'. Help them to identify numbers which have several factors and so appear in more than one 'times table'.

Back at school
Ask the children to choose one of their 'bingo cards' and play 'tables bingo'. Choose two times tables that you are going to test them on, and let them complete a blank bingo card, selecting numbers that they think are likely to be answers!

p83 REARRANGE IT!

Learning outcome
• Understand the principles (not the names) of the commutative, associative and distributive laws.

Lesson context
Demonstrate how to re-group numbers for multiplication using Base Ten materials. Show how re-grouping can make multiplication easier. Work through some simple examples. Talk about the different ways this could be calculated to find the easiest. Let children work in pairs to practise re-grouping.

Setting the homework
Explain that you want them to look at each multiplication and think about the different ways to 're-group' the numbers; they should then select the one that makes the calculation easiest to work out. Talk through the first example with them.

Back at school
Ask children to explain how they re-grouped numbers to complete the calculation. Discuss different options. Ask children to think about how this method could be used to multiply larger numbers. **Answer:** 144, 90, 240, 108, 162, 180.

p84 USING THE GRID!

Learning outcome
• Use informal pencil and paper methods to support, record and explain multiplications.

Lesson context
Demonstrate the grid method of multiplication, pointing out that they must partition the numbers, multiply each part of the number and then total. Explain that this is a multiplication which can be used for numbers of all sizes.

Setting the homework
Tell the children that you want them to do the multiplication using the grid method. Use the example on the homework sheet to ensure that the children understand the basic principle of partitioning.

Back at school
Ask different children to demonstrate to the class how they have reached the answer. More able children could use the method for multiplication of larger numbers.
Answers: 126, 120, 297, 225, 536, 343.

p85 APPROXIMATE FIRST!

Learning outcome
• Develop and refine written methods for TU × U

Lesson context
Explain that you are showing the children a standard method for multiplication. Demonstrate on the board, talking through each stage. Work through examples, encouraging the children to approximate first and then compare the calculated answer with the approximation.

Setting the homework
Point out that there is a worked example on the worksheet.

Back at school
Go through the questions and let children demonstrate their working, comparing different children's approximations. Discuss the methods they used to approximate.
Answers: 1. 413, **2.** 760, **3.** 438, **4.** 756, **5.** 330, **6.** 558.

p86 CHECK IT!

Learning outcomes
• Develop and refine written methods for multiplication and division.
• Use the relationship between multiplication and division.
• Check with the inverse operation.

Lesson context
Revise informal methods for division. Demonstrate how the inverse operation of multiplication can be used to check division calculations. Demonstrate the method for division using multiples of the divisor, stressing that the children should approximate first.

Setting the homework
Explain that you want them to work out the answers on the worksheet using the methods they feel most comfortable with. They must then check their answers by either using a different method or by using an inverse calculation.

Back at school
Ask children to explain their methods and compare the different ways that they have chosen to check their results. Point out that some methods are more appropriate than others for certain calculations and it is important to have a range of strategies available.
Answers; 1. 343, **2.** 19, **3.** 224, **4.** 12.

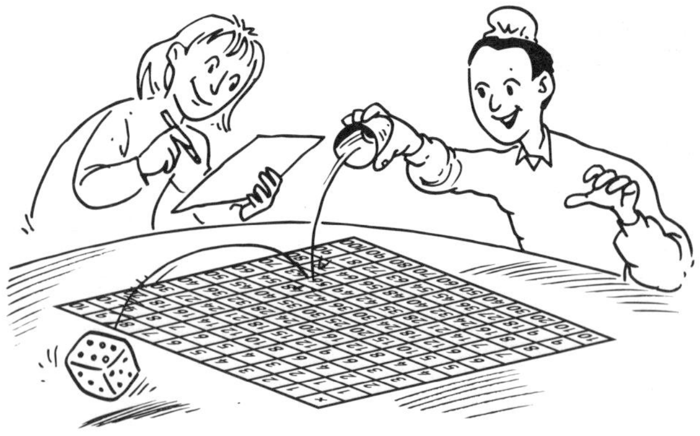

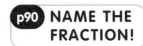

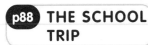

p87 WHAT'S LEFT?

GAMES AND PUZZLES

Learning outcome

- **Find remainders after division.**

Lesson context

Demonstrate how to use a multiplication square for division. Explain that numbers will not always divide exactly, there will often be a 'remainder'. For example to find 27 shared between 5, look on the multiplication square down the '5 times' column for the nearest number below 27, this will give $5 \times 5 = 25$, (remainder 2). Let the children practise finding the remainders after division.

Setting the homework

Tell the children that you want them to practise finding remainders. Talk through how to play 'What's left?'

Back at school

Practise with quick-fire remainder questions, such as $32 \div 6$, $40 \div 3$, $55 \div 8$. Children have to work out the 'remainder' and hold up their digit cards to show the answer.

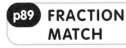

p88 THE SCHOOL TRIP

INVESTIGATION

Learning outcome

- Extend understanding of remainders and know when to round up or down.

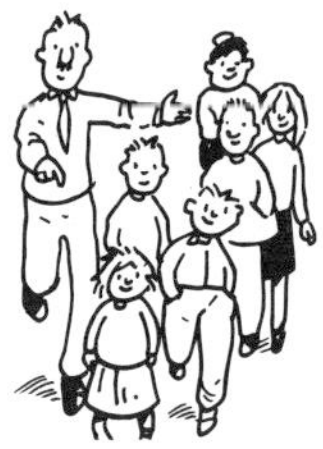

Lesson context

Introduce the work on remainders as described in the previous activity. Discuss the real life situations when remainders are used. Give some examples.

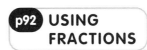

Setting the homework

Explain that you want them to think about remainders in context. They need to think about the investigation and come up with what they consider to be sensible answers. You may prefer to use this activity for the more able pupils.

Back at school

Let children explain their answers and give their reasons. Discuss any difficulties and anomalies! **Answer:** Adults 5, 6, 6, 5. Cheaper coach: 49-seater, £450. Coach for each class: 2×33 seats and 2×49 seats, total cost £500.

p89 FRACTION MATCH

TIMED PRACTICE EXERCISE

Learning outcome

- **Recognise the equivalence of simple fractions.**

Lesson context

Fold equal length strips of coloured card to demonstrate halves, quarters and eighths. Show how these can be put together to make a fraction board, thus demonstrating equal fractions. Let children make their own fraction boards.

Setting the homework

Explain that you want the children to colour in a fraction board and then use it to 'match' equivalent fractions.

Back at school

Go through the answers asking children to talk about the equivalences. Let children investigate other fractions such as thirds and sixths. **Answers:** $\frac{1}{2}=\frac{2}{4}$, $\frac{3}{4}=\frac{6}{8}$, $\frac{1}{3}=\frac{2}{6}$, $\frac{3}{8}=\frac{6}{16}$, $\frac{1}{4}=\frac{2}{8}$.

p90 NAME THE FRACTION!

INVESTIGATION

Learning outcome

- **Recognise simple fractions that are several parts of a whole.**

Lesson context

Demonstrate equivalence of fractions using fraction boards and folded card. Draw a 4×4 grid on the board and shade in fractions asking the children to identify the fractions shaded. Point out that if half is shaded that is the same as $\frac{8}{16}$.

Setting the homework

Explain that you want them to colour parts of various 4×4 grids and to name the fraction that is shaded.

Back at school

Let children demonstrate the different fractions that they have identified, drawing attention to equivalent fractions. Ask the children to draw different grids, such as 3×4 or 5×5 and investigate the different fractions they can shade on these.

p91 MAKE 1

GAMES AND PUZZLES

Learning outcome

- Identify two simple fractions with a total of 1.

Lesson context

Draw a 3×4 grid on the board and invite a child to colour in some of the squares. A second child shades the remaining squares using a different colour. Ask the children to name the fractions shaded and point out that the two fractions make a whole. Ask them to investigate examples of their own.

Setting the homework

Ask the children to name some of the fractions on the 4×4 grid. Explain that there are pairs of fractions that total 1. Ask the children to name each fraction and then find its 'partner' to make 1. There may be more than one answer!

Back at school

Go through the worksheet and discuss any difficulties. Look at the four examples equivalent to $\frac{1}{2}$ and let children talk about which they paired together. When children label fractions, accept cancelled-down answers such as $\frac{8}{16} = \frac{1}{2}$.

Answer: $\frac{5}{16}$, $\frac{1}{4}$, $\frac{3}{8}$, $\frac{8}{16}$, $\frac{12}{16}$, $\frac{2}{4}$, $\frac{9}{16}$, $\frac{4}{8}$, $\frac{7}{16}$, $\frac{1}{2}$, $\frac{11}{16}$, $\frac{10}{16}$.

Pairs: $\frac{5}{16} + \frac{11}{16}$, $\frac{1}{4} + \frac{12}{16}$, $\frac{3}{8} + \frac{10}{16}$, $\frac{8}{16} + \frac{2}{4}$ or $\frac{4}{8} + \frac{1}{2}$, $\frac{9}{16} + \frac{7}{16}$.

p92 USING FRACTIONS

PRACTICE EXERCISE

Learning outcome

- Begin to relate fractions to division and find simple exact fractions such as $\frac{1}{2}$ or $\frac{1}{3}$ of numbers or quantities.

Lesson context

Ask the children questions such as: If you have a bag of 20 sweets and you and your friend have half each, how many will you each have? Explain the relationship between division and fractions using these examples.

Setting the homework

Show the children the worksheet and 'talk through' some of the examples, asking what you have to divide by. For example in question 1, divide by 4 to get $\frac{1}{4}$ of 16.

Back at school

Discuss any difficulties, asking the children to devise their own fraction questions to 'try out' on the rest of the class.

Answer: 1. 4, **2.** 3, **3.** 25, **4.** 6, **5.** 250, **6.** 30, **7.** 5, **8.** 10, **9.** 5, **10.** 25.

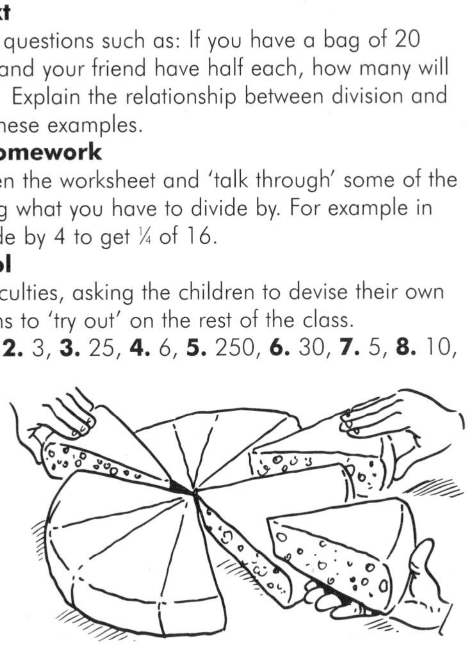

p93 PICTOGRAM

Learning outcome
- Solve a problem by collecting quickly, organising and representing data in tally charts and pictograms – symbol representing 2, 5 10 or 20 units.

Lesson context
Remind children how to produce simple pictograms and give examples. Suggest using an icon to represent more than one unit and talk about how one unit could then be represented.

Setting the homework
Tell the children that they have to produce a pictogram using the data given on the worksheet. Talk about the possible scales that they could use, but remind them that they must use the same scale throughout the pictogram.

Back at school
Ask children to use their pictogram to answer questions such as: *Which is the most popular hobby?* Let the children pose their questions to the rest of the class.

p94 WHICH BOOKS?

Learning outcome
- Solve a problem by collecting quickly, organising and representing data in tables, charts, graphs and diagrams, for example: bar charts – intervals labelled in 2s, 5s, 10s and 20s.

Lesson context
Explain how to organise a tally chart for a book count, discussing the criteria for grouping. Let the children work in groups collecting data from the class books or from the school library. Give each group a different section to 'count'. Collate the data ready for a class graph.

Setting the homework
Explain that you want them to carry out a similar exercise with their own books at home. Talk about designing a tally sheet, pointing out that they do not need a large number of books and that they can include comics, newspapers and other reading material.

Back at school
Let the children talk about the classifications they used for their tally charts, bearing in mind that some children may not have access to as many books as others. Encourage the children to use their data to produce a graph.

p95 GRAPH IT!

Learning outcome
- Solve a problem by collecting quickly, organising and representing data in tables, charts, graphs and diagrams, for example: bar charts – intervals labelled in 2s, 5s, 10s and 20s.

Lesson context
Help the children to produce a bar chart or pictogram from data collected in class. Produce a class graph. Discuss the scale with them and allow them to decide what intervals to use within the confines of their graphs.

Setting the homework
Tell the children they may either use their own data or they can share their friends' data, to produce a bar chart. Stress that you want them to use intervals greater than 1 for their bar chart.

Back at school
Let the children show their bar charts asking them to justify their choice of scales. Discuss with them how they could represent the data in different forms.

p96 TON UP

Learning outcome
- Begin to multiply by 100.

Lesson context
Remind children that when a whole number is multiplied by 10, the digits move one place to the left, and that when a whole number ending in zero is divided by 10 the digits move one place to the right. Then introduce multiplying whole numbers by 100. This time the digits move two places to the left to correspond with the number of zeros. Children work through a series of questions.

Setting the homework
Revise the methods of multiplying and dividing by 10 and then discuss the method of multiplying whole numbers by 100. Explain how they have to be completed. Work through several examples on the board.

Back at school
Run through the children's answers. Revise the relevant rules and ask the children to recite them. Can they suggest what rule would be used for multiplying whole numbers by 1000 and/or dividing whole numbers by 100? **Answers: 1.** 5–500; **2.** 7–700; **3.** 11–1 100; **4.** 14–1 400; **5.** 36–3 600; **6.** 49–4 900; **7.** 75–7 500; **8.** 93–9 300

p97 GIVE ME A SIGN

Learning outcome
- **Use symbols correctly, including less than (<), greater than (>), equals (=).**

Lesson context
Write on the board pairs of numbers for the children to say which one is the larger. Remind the children of the < and > signs. Write up a pair of two-digit numbers and ask which sign should be between them. Repeat using three-digit numbers.

Setting the homework
Run through examples of the type of questions on the sheet. Revise using the correct sign between pairs of numbers. Try examples involving adding, subtracting, multiplying and dividing. Also include doubling and halving.

Back at school
Discuss the answers that have been produced. Ask the children to say the statement aloud using the phrases larger than, smaller than or equal to. In section 3 ask for more than one answer.

Answers: 1. $105 < 115$; $142 > 124$; $656 > 566$; $1702 < 2107$; $9009 < 9900$; double $14 = 28$; $45 > $ double 22. **2.** $52 + 47 < 100$; $155 > 200 - 51$; $253 > 146 + 98$; $194 - 77 < 121$; $50 \div 5 < 44 \div 4$; double $18 = 9 \times 4$; $7 \times 4 > 9 \times 3$; half of $64 < $ double 30.

p98 TOP TENS
MATHS TO SHARE

Learning outcome
- Add three two-digit multiples of 10.

Lesson context
First revise the commutative law for addition. Remind the children of when they regrouped numbers in additions to help with calculations (the associative law). In pairs, children generate 10 pairs of three-digit numbers that they add in the two different orders. They repeat for 10 sets of three two-digit numbers, recording different ways of grouping them to obtain the same answer.

Setting the homework
Remind children about the commutative law for addition, using examples to show that it does not work for subtraction. Remind children also about the associative law for addition, pointing out that it does not apply to subtraction. All questions involve two-digit multiples of 10. Remind them that the answers can be checked by using the inverse operation.

Back at school
Check through the answers, discussing the strategies they used. Ask for examples of how the additions were checked using the inverse operation. Ask the children to repeat definitions of both the commutative and associative laws as they apply to addition. **Possible answers: 1.** $80 + 20 + 30 = 130$;
2. $50 + 40 + 90 = 180$; **3.** $20 + 40 + 50 = 110$;
4. $50 + 70 + 90 = 210$; **5.** $20 + 30 + 50 = 100$;
6. $40 + 30 + 80 = 150$; **7.** $20 + 60 + 40 = 120$;
8. $80 + 70 + 50 = 200$; **9.** $20 + 80 + 70 = 170$;
10. $60 + 30 + 50 = 140$.

p99 MAGIC MACHINES
MATHS TO SHARE

Learning outcome
- Add or subtract the nearest multiple of 10, then adjust.

Lesson context
Remind the children how function machines work. Show them examples of a single and double process function machine. Remind them that an easier way to add or subtract some numbers is to use the nearest multiple of 10 and then adjust. Children work on their own to complete function machines.

Setting the homework
Revise the purpose of function machines and how they work. Work through examples, especially ways of adding and subtracting 9 and adding and subtracting 99.

Back at school
Write the function machines on the board and complete them with the children's help. Discuss in what ways these methods make addition and subtraction easier. Investigate other ways of adding and taking away numbers, linked to other near multiples of 10 such as 31 and 39.
Answers: Add 9: 16–26–25; 24–34–33; 62–72–71;
89–99–98; 114–124–123; 153–163–162.
Subtract 9: 14–4–5; 37–27–28; 73–63–64; 92–82–83;
125–115–116; 164–154–155. **Add 99:** 74–174–173;
98–198–197; 123–223–222; 257–357–356; 406–506–505;
629–729–728. **Subtract 99:** 161–61–62; 193–93–94;
248–148–149; 416–316–317; 677–577–578;
821–721–722.

p100 COLUMN ADDITION
PRACTICE EXERCISE

Learning outcome
- **Develop and refine written methods for column addition.**

Lesson context
Revise long and short column addition with the children. Write the sum vertically, reminding the children that the units must be directly under each other in the units column and similarly for the tens and hundreds. Give an example that involves carrying. Ask children to work through further examples. Recording on squared paper will help them to keep the digits in the correct columns.

Setting the homework
Check the children are proficient at using these methods of column addition.
Go through the stages step by step. Check they know how to set down the questions vertically.

Back at school
Go through the questions carefully using the two different methods and check their answers. Discuss which method they prefer to use and why. Try some other examples that include numbers with zero and three-digit number.
Answers: 1. 231; **2.** 711; **3.** 377; **4.** 521; **5.** 789; **6.** 784;
7. 1132; **8.** 1372.

p101 COLUMN SUBTRACTION
PRACTICE EXERCISE

Learning outcome
- **Develop and refine written methods for column subtraction.**

Lesson context
Remind the children of column subtraction using examples. Invite children to write up subtractions vertically on the board, reminding them to start at the right and work towards the left. Use apparatus to back up explanations.

Setting the homework
Check the children know and understand this method of column subtraction. Check through each of the stages step by step. Ensure they are able to transfer questions from the horizontal to the vertical before working them out. Explain how to use the squared paper to help them to do this.

Back at school
Invite children out to demonstrate their calculations to the rest of the class. Reinforce teaching points especially placing digits in the correct columns. Discuss the different stages in the method. Ask for volunteers to check answers by using the inverse operation. **Answers: 1.** $257 - 98 = 159$;
2. $514 - 76 = 438$; **3.** $632 - 59 = 573$;
4. $754 - 86 = 668$; **5.** Subtract 56 from $497 = 441$;
6. Find the difference between 239 and $872 = 633$.

 **p102 SCHOOL STOCK** INVESTIGATION

Learning outcome
• **Develop and refine written methods for column addition and subtraction;** money calculations.

Lesson context
Write questions like £3.27 + £2.15 and £4.20 + 57p horizontally first and then show how to write them in columns. Stress the numbers and the decimal points must line up vertically under each other. The answers should be labelled with a £ sign. Calculate the answers together, then repeat for money subtractions.

Setting the homework
Revise the stages in the process. Run through the questions on the homework sheet, stressing that answers should be written in pence or labelled with the £ sign. Squared paper should be used to help with setting down. Sets of coins could also be used to help them.

Back at school
Work through the answers. Discuss the strategies they used. Ask them to pick out items they would like to buy themselves. Ask children to show their workings on the board.

Answers: 1a. 70p; **1b.** £1.88; **1c.** £1.17; **1d.** £1.95.
2a. £3.95; **2b.** £3.96;
2c. £4.30; **2d.** £3.90;
2e. £2; **2f.** £3.12.
3a. 30p; **3b.** £1.24;
3c. 99p; **3d.** 25p;
3e. 99p; **3f.** 40p.

 **p103 HOLIDAY TRAVEL** GAMES AND PUZZLES

Learning outcome
• Use addition and subtraction to solve word problems involving numbers in 'real life'.

Lesson context
Go through the word problems and ask children to explain how numbers, signs and symbols can be used to solve the problems. The children then work on their own to complete 'Down on the farm'.

Setting the homework
Revise converting word problems into number form using the correct digits, signs and symbols, using simple examples involving adding and subtracting with both one and two-step processes. Check the children understand the vocabulary used in the questions.

Back at school
Children work through the questions in pairs. One child writes up the numbers, signs and symbols needed to solve the problem, while another child works out the solution. Discuss the strategies used and two examine other suggested methods.

Answers:
1. 217 + 86 = 303; **2.** 345 – 68 = 277;
3. 79 + 105 + 263 = 420;
4. 358 – 179 = 179; **5.** 359 – 109 + 54 = 304;
6. 30 + 31 + 31 – 17 = 75; **7.** 300 – 150 – 75 = 75;
8. adults 569; children 97; total 666.

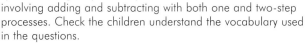 **p104 MATCHING MEASURES** PRACTICE EXERCISE

Learning outcome
• **Know and use the relationships between familiar units of capacity.**

Lesson context
Revise the key units used to measure length and mass and then introduce capacity as the amount something holds in litres and millilitres. Show the children the measuring containers. Ask questions that require them to convert from one to the other. Introduce children to the key fractional parts of a litre and the decimal equivalents. Children then work on a metric capacity task.

Setting the homework
Revise the work above. Work through some examples where children have to order capacity amounts starting with the smallest involving fractions and decimals.

Back at school
Ask different children to give their answers. Ask children to find real life examples of capacity measurements at home and see if they can write the amounts in a different way.

Answers: 1a. 5 000ml; **1b.** 3 500ml; **1c.** 6 250ml;
1d. 9 750ml; **1e.** 4 100ml; **1f.** 2½l; **1g.** 7l; **1h.** 3¾l;
1i. 8⅒l; **1j.** 12¼l. **2a.** 1l 200ml; **2b.** 5l 400ml;
2c. 3l 245ml; **2d.** 6l 5ml; **2e.** 4 350ml; **2f.** 8 615ml;
2g. 7 400ml; **2h.** 9700ml. **3a.** 300ml, 500ml,
600ml; **3b.** 750ml, 800ml, 900ml; **3c.** 1 250ml,
1 900ml, 2 000ml; **3d.** 3 700ml, 3 750ml, 3 900ml;
3e. 5 250ml, 5 450ml, 5 900ml; **3f.** 4 100ml,
4 250ml, 4 500ml.

 **p105 CAPACITY QUIZ** MATHS TO SHARE

Learning outcome

• Record estimates and readings from scales to a suitable degree of accuracy.

Lesson context
Explain that they are going to find out how much a variety of containers will hold. Tell them to estimate first and then choose which container and measuring cylinder to use to measure the amount. They record their results on the prepared charts. The children work in groups on practical tasks.

Setting the homework
The children could use measuring jugs or a plastic litre bottle, marked with a scale on the side. Talk about the importance of honest estimating first and then accurate measurement. Stress that spillage will spoil their results.

Back at school
Discuss the containers they investigated and the apparatus they used for the measurements. Were the containers easy to fill up? Were the scales easy to read? What did they find the most difficult part of measuring? How close were their estimates? Did it get better with practice?

p106 MIRROR IMAGE

Learning outcome
• Sketch the reflection of a simple shape.

Lesson context
Show the children how to mark the axes on grid-paper. Tell them that the vertical line is the mirror line, then begin a pattern by colouring in corresponding squares on either side of the line. Show them how to plot corresponding squares. Children work individually to create their own symmetry patterns.

Setting the homework
Tell the children they are going to make coloured symmetrical patterns that correspond with the ones that are already shown on one part of the grid. Check they understand the colour coding system. Have squared paper available to take home for those able to produce their own coloured patterns later.

Back in school
Use large-squared paper mounted on the board to check out the patterns. Fill in the coloured squares given on the sheet and invite children to complete the symmetrical patterns by colouring squares on the other side of the line or lines. Use a large mirror to check that the shapes have been drawn accurately. Revise how the positions were plotted.

p107 LETTER LAND

Learning outcome
• Recognise positions and directions and use co-ordinates.

Lesson context
See details from the previous lesson as the children will again be plotting co-ordinates in squares rather than points where two lines cross.

Setting the homework
Point out that this time the squares across are labelled with letters and the squares going up are labelled with numbers. Remind the children that once identified, the whole square should be coloured in. Point out that the fourth square has been left blank so that they can make up their own letter shape. They should write down the co-ordinates and then ask a friend to make it up.

Back at school
Check through grids 1, 2 and 3 to make sure that they have coloured in the correct letter shapes: 1 – F, 2 – S and 3 – Y. Move on to children's own letter shapes. Put some large squared paper up on the board. Ask one child to call out their co-ordinates while another child colours in the correct squares on the squared paper.

p108 SHAPE UP

Learning outcome
• Recognise positions and directions and use co-ordinates.

Lesson context
The children are going to plot positions where two lines on the grid cross. Using the large piece of grid paper, show children how to locate different positions e.g. (6,3), (2,1), (5,4) and (3,2). Mark the point where the two lines cross with a dot or small cross. Give them the task 'Plotting shapes'. They should mark the points in the order given, joining them up with straight lines using their ruler.

Setting the homework
Revise the information given above. Children may prefer to use a coloured pencil so that lines stand out clearly. Tell children that the fourth grid should be used to make up their own 2-D shape, different from the ones already featured.

Back at school
Go through the shapes on the homework sheet. Ask children to plot the shapes on a large piece of squared paper already marked with the grids. Now ask for the grid references for the shapes they have made themselves for others to plot on the large grid paper. **Answers: 1.** pentagon; **2.** parallelogram; **3.** hexagon.

p109 RIGHT DIRECTIONS

Learning outcome
• Make and measure clockwise and anticlockwise turns.
• Use the eight compass directions.

Lesson context
Divide a large circle into four quadrants and mark out an eight-point compass rose. Discuss the eight points of the compass and clockwise and anticlockwise turns. Children then make their own eight-point compass rose using circles of coloured paper. They should fold to make a right angle and then fold again to make eighths. They then use their compass to work on 'Take it in turns'.

Setting the homework
Revise the eight points of the compass and the meaning of the words clockwise and anticlockwise. Check the children understand the terms quarter of a turn, half a turn, three-quarters of a turn and a full or complete turn. Ensure the children understand that each question on the worksheet starts at a stated position.

Back at school
Have some blank compass roses on the board ready so you can invite children out to show you how they moved in the given direction and what direction they finally reached. Revise key-words. **Answers: 1.** South; **2.** East; **3.** East; **4.** North; **5.** East; **6.** South East; **7.** North East; **8.** South West.

p110 ANGLE CHALLENGE
INVESTIGATION

Learning outcome
- Begin to know that angles are measured in degrees. Start to order a set of angles less than 180°.

Lesson context
Divide a large circle into four quadrants with a vertical and horizontal line through the centre. Mark in a quarter turn clockwise with a coloured pen. Tell the children this is one right angle or 90°. Repeat for a half, then for a three-quarter turn and finally for a complete turn. Children then work on their own to complete an activity similar to the homework.

Setting the homework
Show the children practically how to make the right angle checker by folding a small piece of squared paper twice. They can use it to see if angles are less than, about the same as or more than 90°. Point out this should help them sort them sort the angles into order of size starting with the smallest.

Back at school
Sort the angles on the sheet first into three groups: less than 90°, about 90° and more than 90°. Get the children to sort the angles into order of size starting with the smallest. Introduce new words they will need later on: acute angles, right angles, obtuse angles and straight line angles.

Answers: Order of size of angles starting with the smallest: 7, 1, 4, 6, 2, 3, 5, 8.

p111 MULTIPLE SORT
INVESTIGATION

Learning outcomes
- Recognise and extend number sequences formed by counting from any number in steps of constant size.
- Recognise multiples of 2, 3, 4, 5 and 10.

Lesson context
Remind children that multiples of 10 will always have a 0 as the last digit and that multiples of 5 end in 5 or 0. They also know that multiples of 2 are all even numbers. Explain that a test for recognising a multiple of 4 is that when the number is halved it is even. Let children practise identifying multiples of 2, 4, 5 and 10.

Setting the homework
This is an investigation with numbers up to 30, to find which are multiples of different numbers. Give some random examples and work them through with the class. Point out that they do not have to tackle them in order but they can look for the ones they know first.

Back at school
Discuss which numbers are multiples of more than one number. You may want to introduce the terms factor and prime numbers. Prepare a class chart with numbers up to 50 (or 100) as an extension exercise.

p112 THE RULE OF THREE
INVESTIGATION

Learning outcomes
- Recognise and extend number sequences formed by counting from any number in steps of constant size.
- Recognise multiples of 2, 3, 4, 5 and 10.

Lesson context
Write the 3 times table on the board and ask the children to add together the digits in each answer. Point out that that there is a pattern. Explain that a way of identifying a multiple of 3 is when all the digits in the answer add up to 3, 6 or 9. Let the children play a game generating numbers and then deciding whether they are multiples of 3.

Setting the homework
This homework is an investigation to reinforce the information they have been given in the lesson. Explain that you want them to continue the investigation into the sum of the digits of the answers in the 3 times table. Remind them that they can find the higher multiples by using the 100 square or by counting on.

Back at school
Discuss the outcomes of the investigation, letting the children tell you about the higher numbers that they investigated. Give some larger numbers to see if they can work out whether they are multiples of 3 or not, by adding the digits.

p113 IS IT A MULTIPLE?
TIMED PRACTICE EXERCISE

Learning outcomes
- **Know by heart multiplication facts for 2, 3, 4, 5 and 10 times tables.**
- Begin to know multiplication facts for 6, 7, 8 and 9 times tables.
- Use doubling and halving, starting from known facts.

Lesson context
Discuss the different strategies they can use to complete a multiplication square. Let them work with a partner to complete a multiplication square, and then demonstrate how they can use it to answer multiplication and division questions.

Setting the homework
Explain that you want the children to complete a multiplication grid using the strategies that they know. Tell them to try to do it as quickly as possible and then do the 'sums' at the end. This is an activity that children can do repeatedly, trying to beat their best time.

Back at school
Let the children use the multiplication squares to answer some quick-fire multiplication questions. Explain how to use it for division questions and give examples of these.

p114 **TABLES SQUARE**    TIMED PRACTICE EXERCISE

Learning outcomes

- **Know by heart multiplication facts for 2, 3, 4, 5 and 10 times tables.**
- Begin to know multiplication facts for 6, 7, 8 and 9 times tables.
- Use doubling and halving, starting from known facts.

Lesson context

Discuss the different strategies they can use to complete a multiplication square. Let them work with a partner to complete a multiplication square, and then demonstrate how they can use it to answer multiplication and division questions.

Setting the homework

Explain that you want the children to complete a multiplication grid using the strategies that they know. Tell them to try to do it as quickly as possible and then do the 'sums' at the end. This is an activity that children can do repeatedly, trying to beat their best time.

Back at school

Let the children use the multiplication squares to answer some quick-fire multiplication questions. Explain how to use it for division questions and give examples of these.

p115 DOUBLE IT!   MATHS TO SHARE

Learning outcome

- Use doubling and halving, starting from known facts.

Lesson context

Demonstrate how they can find doubles of larger numbers by doubling the tens, doubling the units and then adding the two together. Let the children generate numbers using dice and then work out their doubles.

Setting the homework

Tell the children that, after a 'warm up' on the worksheet, you want them to work with their helpers practising doubling and halving. Remind them to partition the number and use the doubling facts that they already know.

Back at school

Let children challenge each other to double and halve numbers, aiming for five correct answers in a row.

p116 TIMES IT!   TIMED PRACTICE EXERCISE

Learning outcome

- Develop and refine written methods for TU × U, TU ÷ U.

Lesson context

Demonstrate column multiplication on the board. Stress that the children should approximate first and then calculate. Demonstrate the extended method for column multiplication and then explain that this can be shortened by first multiplying the least significant digit and then 'carrying' the tens.

Setting the homework

Go through the worked example using both the extended and the shortened methods. Tell the children that you want them to try to use the shortened method and to time themselves on the activity.

Back at school

Go through the answers and discuss any difficulties. Ask the children about the accuracy of their approximations. Point out that this is an important way to check their answers for reasonableness. **Answers: 1.** 399, **2.** 294, **3.** 504, **4.** 376, **5.** 539, **6.** 420.

p117 DIVIDE IT!   PRACTICE EXERCISE

Learning outcome

- Develop and refine written methods for TU × U, TU ÷ U.

Lesson context

Write a division on the board and ask the children to approximate the answer. Now demonstrate column division, explaining each step. Go though some examples. Make sure that the children fully understand how to check their answer for reasonableness using approximation.

Setting the homework

Tell the children that this is a practise exercise as a follow-up to what they have been doing in class. Remind them to approximate first and then calculate.

Back at school

Invite children to work through the examples, demonstrating to the class. Discuss any difficulties. **Answers: 1.** 15, **2.** 22, **3.** 23, **4.** 24, **5.** 19, **6.** 12.

p118 REMAINDERS   PRACTICE EXERCISE

Learning outcomes

- Develop and refine written methods for TU × U, TU ÷ U.
- **Find remainders after division.**

Lesson context

Demonstrate an example of column division with a remainder on the board. Let the children work through some similar examples making sure that they set the sums out correctly.

Setting the homework

Tell the children that this exercise is to give more practice in column division, following up what they have been doing in class with remainders. Remind them to approximate first and then calculate.

Back at school

Invite children to work through the examples, demonstrating to the class and 'talking through' what they have done. Discuss any difficulties and ask the children what the 'biggest remainder' can be for a division sum.

Answers: 1. 14r1, **2.** 12r1, **3.** 16r2, **4.** 22r2, **5.** 13, **6.** 16r3, **Extension a.** 14 sweets, 2 left, **b.** 14 crayons, none left.

p119 PARTY SHOPPING GAME   GAMES AND PUZZLES

Learning outcome

- **Find remainders after division.**

Lesson context

The children need to remember that there are 100 pence in a pound when dividing money. Demonstrate some division problems involving money, pointing out that sometimes there will be remainders.

Setting the homework

Show the children the homework sheet and explain how the game works. Make sure that they understand that they have to 'buy' the items on the shopping list. They can save money from one 'go' to the next, but they need to keep a careful record of the transactions.

Back at school

Ask children to talk about any difficulties they encountered. Let the children play the game in small groups, then encourage them to think up their own shopping lists and price lists to invent their own version of the game.

p120 FRACTION DECIMAL MATCH
PRACTICE EXERCISE

Learning outcomes
- Recognise the equivalence between the decimal and fraction forms of one half, one quarter and tenths.

Lesson context
Using a number stick marked in tenths, talk to the children about decimal and fraction equivalences that they already know, such as ½ is equal to 0.5. Demonstrate, for example, that ³⁄₁₀ can be represented as 0.3. Let children practise matching decimals to fractions.

Setting the homework
Show the children the homework sheet and explain that you want them to complete the comparison grid first to help them match the fractions to their decimal equivalents. Suggest they label the fractions ¹⁄₁₀, ²⁄₁₀ etc. Remind them to look carefully at the mixed fractions and give some examples of these.

Back at school
Go through the homework sheet and discuss any difficulties the children experienced. Look particularly at ¹⁄₁₀, 1 ¹⁄₁₀ and 10 ¹⁄₁₀ and their equivalents, making sure that the children understand the place value involved with each of these. Ask the children to order the decimals according to size, either as a group or individually. **Answers:** ¼–0.25, ³⁄₁₀=0.3, ¹⁄₁₀=0.1, 1½=1.5, 1¹⁄₁₀=1.1, ⁷⁄₁₀=0.7, 3½=3.5, 10¹⁄₁₀=10.1, ¾=0.75, ⁹⁄₁₀=0.9.

p121 HIGHEST / LOWEST
GAMES AND PUZZLES

Learning outcome
- Understand decimal notation, and use it in context.

Lesson context
Remind children about decimals as tenths. Introduce hundredths and relate this to money. Explain the importance of the most significant digit when ordering a set of decimal numbers, stressing that zeros on the right of a decimal number are not significant. Give the children some activities involving ordering decimal numbers.

Setting the homework
Explain the game 'Highest/lowest' and have a 'practice run'. Ask a child to select three digits and then talk to the class about the different numbers that can be made. Order these numbers, stressing that the decimal point must be included.

Back at school
Let the children play 'Highest/lowest' as a class, using the board. Pay particular attention to numbers with 0 included, you cannot accept 01.2 or .012, but do accept 0.12.

p122 DECIMAL HUNT
INVESTIGATION

Learning outcome
- Understand decimal notation, and use it in context.

Lesson context
Talk to the children about recording amounts of money in decimal form. Explain how pence can be written as decimals of a pound. Let the children practise converting amounts of money from pence to decimals of a pound and vice versa.

Setting the homework
For this investigation you want them to look at items in the home, especially the kitchen and bathroom. Explain that they are looking for decimals. They may find prices on items, but quantities are not often shown in decimal notation on household goods. Ask the children to list quantities on the chart that they can record in a decimal form. Explain the example that is shown on the worksheet.

Back at school
Ask the children to share the quantities that they have found and their decimal equivalents. Discuss the different amounts and make a class chart on the board.

p123 MONEY ADDS
PRACTICE EXERCISE

Learning outcome
- **Develop and refine written methods for column addition and subtraction of two whole numbers less than 1000, and addition of more than two such numbers;** money calculations.

Lesson context
Demonstrate how to set out column addition sums involving money. Stress the importance of positioning the digits carefully with the decimal points 'in line'. Go through some examples and then give the children some amounts to add up; encouraging them to practise writing the sums vertically.

Setting the homework
Show the children the worksheet and 'talk through' the first example. Explain that you want them to try to do this on their own, but if they need help to ask their Helper.

Back at school
Go through the answers and discuss any difficulties. Ask children to demonstrate how they did the last three questions looking for any children who made errors because they wrote the sum down incorrectly.

Answers:
1. £9.69, **2.** £9.24, **3.** £12.63, **4.** £7.45, **5.** £3.35, **6.** £12.20, **7.** £13.30, **8.** £10.70, **9.** £7.18.

p124 SHOPPING CHECK!
GAMES AND PUZZLES

Learning outcome
- **Develop and refine written methods for column addition and subtraction of two whole numbers less than 1000, and addition of more than two such numbers;** money calculations.

Lesson context
Demonstrate how to set out and calculate money additions. Stress the importance of positioning the digits correctly and the decimal point. Remind them to add the pence first.

Setting the homework
Explain the activity to the children. Have some spare supermarket receipts for any children who may not have one at home. Demonstrate how to divide the receipt into sections and add the amounts separately, making sure they realise how they can check the total.

Back at school
Talk about any difficulties. Go through different methods for checking addition, such as adding amounts in reverse and splitting the amounts into sections and totalling them.

Learning outcome

- Solve a problem by collecting quickly, organising and interpreting data in tables, charts, graphs and diagrams.

Lesson context

Demonstrate how a Venn diagram can be used. You may like to use data that has been collected for the work on Carroll diagrams, alternatively sort numbers using criteria such as even numbers and multiples of 5. Let the children investigate a variety of Venn diagrams.

Setting the homework

Show the class the worksheet and explain that you want them to complete the Venn diagram. Remind them that they can use a mirror to check whether the letters have reflective symmetry. Talk about what criteria will apply to the overlapping section of the diagram.

Back at school

Draw a large Venn diagram on the board and let children come out and position the letters where they think they should be. Discuss any difficulties and let children explain their decisions.

p125 TV TIMES
MATHS TO SHARE

Learning outcome

- Read simple timetables.

Lesson context

Talk to the children about when we use timetables, calendars and various other tables and let them investigate a selection of these. Point out that they are used as sources of information. Let the children use calendars or timetables to find information.

Setting the homework

Talk to the children about time. Show a TV page from a newspaper and discuss the start and finish times of various programmes. Help children to work out the length of each programme. Run through the sheet, explaining what they have to do.

Back at school

Let children share their programme plans. Design a class list and talk about whether programmes should follow on, or whether there should be a gap.

p126 CARROLL SORT
INVESTIGATION

Learning outcome

- Solve a problem by collecting quickly, organising and interpreting data in tables, charts, graphs and diagrams.

Lesson context

Demonstrate the use of a Carroll diagram on the board, asking children to enter information in the boxes. Explain that with a Carroll diagram the headings are, for example, 'blue eyes' and 'not blue eyes'.

Setting the homework

Show the children the worksheet and explain that you want them to collect data about some of their friends and family. They must then think very carefully about which heading to choose for their Carroll diagram.

Back at school

Let children show their diagrams and compare the data. Discuss how the data could be collated to make one big Carroll Diagram.

p128 HOUSES AND HOMES
MATHS TO SHARE

Learning outcome

- Solve a problem by collecting quickly, organising and interpreting data in tables, charts, graphs and diagrams.

Lesson context

Demonstrate a computer database programme such as 'Pinpoint'. Collect some data in school and input it to the database, talking to the children about different designs for a database. Explain that it is helpful to prepare a questionnaire to collect data for a database.

Setting the homework

Show the children the worksheet and explain that they need to try to complete the questionnaire accurately and then input the data to the computer. Talk through the questions, making sure that the children understand the questions.

Back at school

Prepare the database on the computer and allow the children to input their data working in pairs at the computer. Ask the children to think of questions they can ask such as: *How many children have a garage at home?*

Abacus charts

You will need: coloured pencils, spare sheets of paper.

These charts will help you with four-digit numbers.

In these examples, the digits and the words are written for you.

- Draw the correct number of beads on each abacus chart with a coloured pencil.

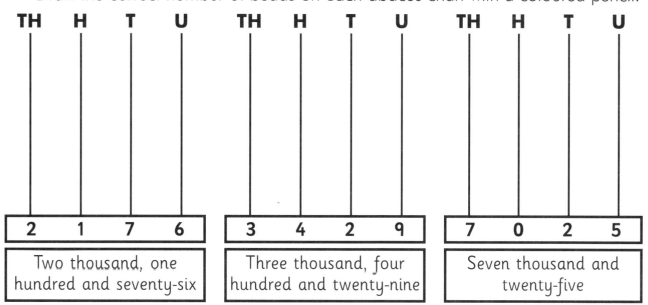

TH	H	T	U
2	1	7	6

Two thousand, one hundred and seventy-six

TH	H	T	U
3	4	2	9

Three thousand, four hundred and twenty-nine

TH	H	T	U
7	0	2	5

Seven thousand and twenty-five

In these questions, the beads are in place.

- Write the numbers shown in both digits and words.

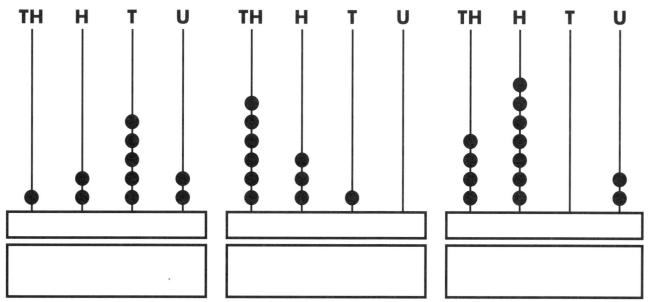

Dear Helper,

Abacus charts will help your child improve their understanding of four-digit numbers up to 9 999. Let your child complete the worksheet alone and then discuss the answers. Encourage your child to say the numbers as they are recorded. Zero columns should be left blank, but the digit 0 must be written under the correct column. Your child can practise other examples of these charts using numbers up to 9 999, using either the back or this sheet or spare paper.

Name:

Add or subtract

You will need: red and blue counters big enough to cover the circles.

This is a game for two players.

- One player uses the red counters, the other uses blue.

- Decide who will start first, then take it in turns.

- When you find a correct answer on the right-hand side, cover it with your counter.

- The first one to cover all the circles is the winner.

Add or subtract 100 to/from these numbers:

To cover these answer circles:

72	123	245
1532	1431	6174
	196	7141

7041	1632	172
145	296	6074
	23	1331

Add or subtract 1000 to/from these numbers:

235	1177	646
	1251	4376
5300	15	4942

To cover these answer circles:

3376	251	4300
	3942	1015
1646	1235	177

Dear Helper,

Play this game with your child. Before starting the game, discuss what will happen when adding/subtracting if the digits 9 or 0 occur in the hundreds column. Ask: *How will this affect the other digits in the number?* Encourage them to work out the answer mentally if they can. Discuss what strategies they are using to do this. If they are uncertain, they should use a paper and pencil method to check the answer.

Jumping kangaroo

Your pet kangaroo is not only clever at jumping, he is also good at maths!

He always jumps in regular patterns, forwards or backwards. Some days he jumps on or back in hundreds. Here are some examples:

347   447   547   647   1236   1136   1036   936   836

Some days he jumps on or back in thousands. Here are some examples:

6843   5843   4843   3483   2307   3307   4307   5307

- Keep up with your pet kangaroo by writing the next three numbers in these jumping patterns.

Remember you will be counting on or back in hundreds or thousands.

1	977	877	777	677	_____	_____	_____
2	3121	4121	5121	6121	_____	_____	_____
3	1520	2520	3520	4520	_____	_____	_____
4	6125	6225	6325	6425	_____	_____	_____
5	9117	8117	7117	6117	_____	_____	_____
6	5646	5746	5846	5946	_____	_____	_____

- Some days he is so clever he even mixes up the two kinds of pattern in the same journey. What is the jumping rule in each of these sequences?

7   324   424   1424   1524   2524   _____ _____ _____

The rule is _____

8   8545   7545   7445   6445   6345   _____ _____ _____

The rule is _____

- Give the next three numbers in each one.

- Now make up your own sequences for your jumping kangaroo using hundreds and thousands.

CALCULATIONS

MENTAL CALCULATIONS + AND –

Number splits

- Practise adding these numbers using the partition method. In this method, the hundreds, tens and units are added separately before being joined together.

Here are two examples:

1 215 + 64 = 215
         64
      ─────
        200
         70
          9
      ─────
        279

2 427 + 315 = 427
          315
       ─────
         700
          30
          12
       ─────
         742

Now try adding these numbers using this method.

1 529 + 45

2 814 + 94

3 397+ 33

4 251 + 495

5 406 + 158

6 579 + 206

Dear Helper,

After your child has worked through the questions independently, ask them to explain the partition method of addition to you. This will demonstrate their understanding. Before working out, ensure the question is changed from the horizontal to the vertical as shown above, as this will help with the more formal methods of column addition that your child will meet later.

PHOTOCOPIABLE

In a word

In maths, different words are used to mean the same operation. For example, 'plus' and 'find the total of,' both mean that numbers have to be added, while 'subtract' and 'minus' tell us that one number has to be taken away from another.

- Find the missing numbers in these questions: the answers are under 20.

- Carefully check the meanings of the words to make sure you carry out the correct operation.

- Write down how you worked out the answer under each question, showing the mathematical signs clearly.

1 This number is the total of 7, 5 and 3.

2 This number is the difference between 17 and 9.

3 Decrease 14 by 5.

4 What do 9, 7 and 3 make altogether?

5 Find the sum of 5, 6 and 9.

6 Eleven plus four.

7 Subtract 8 from 17.

8 Eighteen minus 12.

9 What number is 7 more than 10?

10 Find this number. It is larger than 10 but less than 15. It is a prime number. Its digits add up to four.

Dear Helper,

Understanding the meaning of key vocabulary is important in all mathematical operations. Revise the words used to indicate either addition or subtraction with your child before starting the activity. Discuss any other words in common use for the same operations. Challenge your child to make up their own 'word-based' problems for numbers up to 20, using a dice to generate the numbers. They could record them on the back of this sheet.

Name:

All change

You will need: a set of coins;

Section A

Use the least number of coins to make up these money amounts:

27p _____

84p _____

49p _____

£1.17 _____

61p _____

£3.56 _____

Section B

Use the coins in your collection to solve these money problems. You will need to carry out at least two steps to get the answer. Show your working out alongside each question or use the back of the sheet.

1 Sanjay has a 50p coin, a 20p coin and a 5p coin. How much more money does he need to buy a notebook costing 85p?

2 If you spent 24p on sweets and 35p on a comic, how much change would you have from £1.00?

3 Mary has two 50p coins and a 10p coin. Ann has four 20p coins and a 5p coin. How much more money does Mary have?

4 Ahmed empties his moneybox. He has five 10p coins, three 5p coins and two 2p coins. He loses one of each kind of these coins. How much does he have left?

5 If you started with £2.48, how much money would you have left if you bought a 55p magazine and four pencils costing 12p each?

Dear Helper,

If possible, please provide your child with four or five of each of the lower denomination coins so they have plenty to handle. Sit with them and check the correct coins have been used when they have finished each question. Discuss possible alternatives when they are making up amounts. For example, 32p could be made up as 20p, 10p and 2p, or 10p, 10p, 10p and 2p, or 20p, 5p, 5p and 2p.

Odd and even

You will need: a coloured pencil.

- Find the answers to these questions by adding or subtracting.

- Now find the answer inside a square on the grid.

If it is an odd number colour it in.

If it is an even number leave it blank.

12	26	4	30	78	39	43	71
2	21	9	41	22	51	50	76
16	35	8	29	6	41	65	27
24	19	13	7	10	17	26	40
30	48	16	44	36	5	33	3

9 + 3 40 − 38 6 + 10 20 + 4 50 − 20	19 + 7 30 − 9 15 + 20 20 − 1 18 + 30	24 − 8 7 + 6 50 − 42 25 − 16 16 − 12	70 − 40 31 + 10 40 − 11 29 − 22 14 + 30
28 + 50 63 − 41 40 − 34 90 − 80 13 + 23	50 − 11 17 + 34 57 − 16 9 + 8 75 − 70	17 + 16 52 − 26 47 + 18 100 − 50 17 + 26	94 − 23 51 + 25 14 + 13 100 − 60 52 − 49

Which two letters have you shaded in on the grid? _____

Use the answers you have found to complete these statements.

Even + even equals _____   Even − even equals _____

Odd + odd equals _____   Odd − odd equals _____

Even + odd equals _____   Even − odd equals _____

Odd + even equals _____   Odd − even equals _____

Dear Helper,

Working out the questions to complete the grid will produce some evidence to complete the statements, but encourage your child to try out other examples to prove the rules. Start with mental addition or subtraction of single-digit numbers and then progress to two-digit, and even three-digit numbers. Discuss examples in real life, such as the use of odd and even numbers in most street numbering systems.

Name:

Different order

You will need: to collect data from around the home that will give you four addition sums involving at least three two-digit numbers.

For example, count the number of packets and tins on different shelves in a cupboard. You could count up the number of books on each shelf of a bookcase. Try counting up the number of CDs owned by different people in the family, or the number of plants or flowers in different areas of the garden.

Once the numbers have been collected, add them up in several different orders. Does changing the order of the numbers change the result?

Discuss what you have found out with your Helper.

Example 1:

Example 2:

Example 3:

Example 4:

Dear Helper,

Help your child find suitable situations at home where they can collect the data they need for this task. In this activity, your child will be working on the 'Commutative Law' for addition: the fact that the same numbers can be added in any order and still produce the same answer. For example, 13 + 12 = 25 and 12 + 13 = 25. Once the practical work has been completed, reinforce this maths law with pencil and paper examples, including the use of some three-digit numbers.

Name:

Measure up

You will need: a 30cm ruler marked in both centimetres and millimetres.

• Use the ruler to measure the different parts of this house picture.

• Record your results in the boxes at the bottom of the sheet.

• Write your answers in at least two different ways. For example: 5cm as 50mm or 3cm 4mm as 3.4cm.

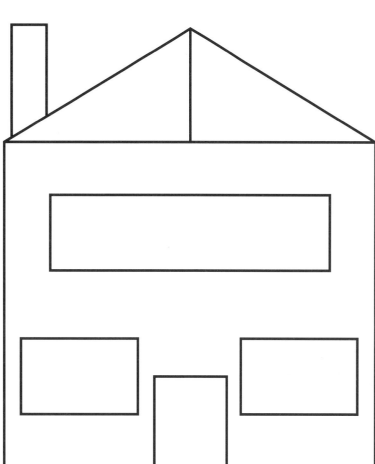

Width of house	Height of house	Height of roof	Height of door

Width of downstairs window	Width of upstairs window	Width of chimney	Height of chimney

Dear Helper,

Please ensure that your child is able to read and understand the divisions on the 30cm ruler. Check the measurements yourself to make sure their answers are precise. Help them to convert measurements into different units, writing them in more than two ways if possible e.g. 5cm = 50mm = 0.05m.

Name:

Long distance

You will need: a tape measure that is at least one metre long.

- Look at the scale on the tape measure and make sure you can read it. If not, ask your Helper to explain.

- Find eight, straight-sided objects in the home that you can measure easily and safely, such as a table, low cupboard, piece of carpet, or window/door frame.

- Write down their names in the first column of the table below and then fill in the other sections. Be honest about your estimating and measure to the nearest centimetre.

Object	More/less/same as a metre	Estimate	Measure
1			
2			
3			
4			
5			
6			
7			
8			

Dear Helper,

The purpose of this activity is to help your child measure accurately with a metre tape measure. Check that they understand the sub-divisions of a metre marked on the tape. Encourage honest estimating, making a sensible guess first, even if their first attempts are a long way out. Help your child to round up or down to the nearest centimetre. Remember 10mm = 1cm and measurements up to and including 4mm will go back to the previous centimetre, whilst 5mm or more will go up to the next centimetre.

Cover up

You will need: a 30cm ruler and some squared paper.

- Find the areas and perimeters of the squares and rectangles below.

Remember, area is the amount of surface in a shape and is measured in square centimetres (cm^2). The perimeter is the distance around the outside of the shape and is measured in centimetres (cm).

Marks have been placed on the shapes to help you draw the squares if you need to, but try to look for quicker ways of working out the answers.

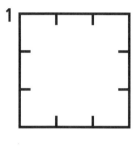

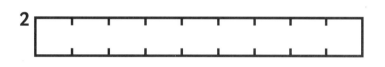

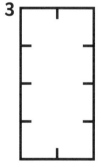

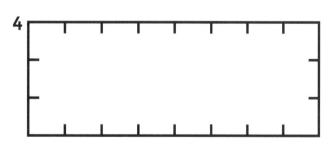

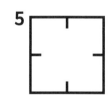

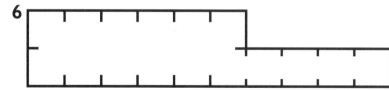

Dear Helper,

Check first that your child can tell you the difference between area (surface size) and perimeter (distance round) and knows the correct units to record the answers. Marks are provided to help you draw in grid lines if your child gets stuck and wants to count the squares. Encourage them to avoid drawing in individual squares and counting them up to find the answer. Look for quick ways of working so they begin to understand that the area of a square or rectangle can be found by using the formula length × breadth and perimeter can be found by adding up the lengths of the sides.

Name:

All shapes and sizes

- Look carefully around the inside and outside of your home for two-dimensional (2-D) flat shapes that you find around you.

- Draw as many examples as you can of these 2-D shapes, using a ruler for any straight lines.

- Start with squares, rectangles, circles and triangles, but also look out for pentagons (five-sided), hexagons (six-sided), heptagons (seven-sided) and octagons (eight-sided) shapes.

- Try to give examples of both regular and irregular shapes.

You can continue on the back of this sheet.

Dear Helper,

Let your child find, draw and label the shapes independently, to see how many they remember. Work together to check through their answers. Talk about key words such as sides, angles and vertices (corners). Can your child show you these parts? Discuss the difference between regular and irregular 2-D shapes. In regular shapes all the sides and angles are equal, while in irregular shapes they are not. Can your child show you examples of both types?

Name:

In the net

You will need: a pencil, ruler, sticky tape or glue and some triangular grid paper.

One of the nets of a tetrahedron, or triangular based pyramid, is drawn on the triangular grid paper below.

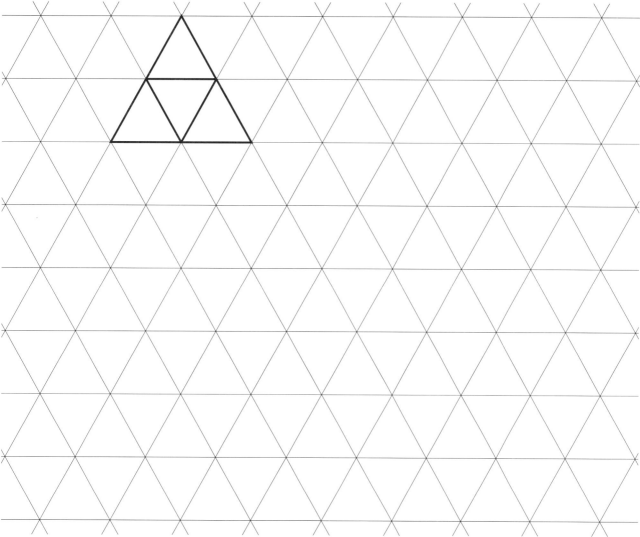

- Cut out the shape carefully so that you can make it up into the 3-D shape. Include gluing flaps to help the shape fit together neatly.

- Now draw other nets of the same shape on the grid paper, again with gluing flaps. Cut them out and see if they fit together successfully.

How many different nets of a tetrahedron can you find?

| Dear Helper, |

First discuss the concept of a net with your child. That is, how a 3-D shape looks in its flattened-out state before it is constructed. Talk about the importance of gluing flaps that go on the inside of the shape to hide the joins and hold the shape together. Once a successful tetrahedron has been made, help your child find the number and shape of its faces, how many edges it has, and how many vertices or corners there are.

Name:

Double trouble

You will need: some large coloured counters or coins.

This is a game for two people.

- Toss a coin to see who will start first, then take it in turns.

- In the first half, double the number on the ball at the footballer's feet, and cover up the correct answer on the ball in the goal-mouth with a counter or coin.

- In the second half, halve the number on the ball and again cover the correct answer on the ball in the goal-mouth. The first one to cover all the correct answers is the winner.

First half: Remember to double

Second half: Remember to halve

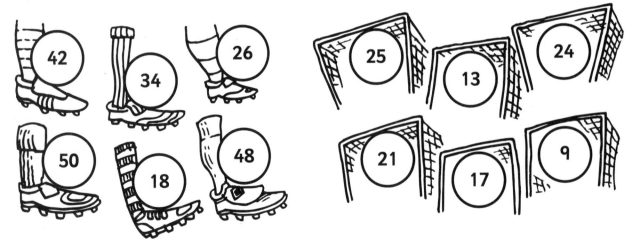

Dear Helper,

Your child will have been taught that halving is the inverse of doubling. For example, if half of 40 is 20 then double 20 is 40. Practise some examples of this before they start work on the sheet. Many children find it easier to partition numbers when doubling and halving. For example, double 32 is double 30, then double 2 = 60 + 4 = 64. Half of 48 is half of 40 then half of eight = 20 + 4 = 24.

Times three

You will need: some coloured pencils.

- Look carefully at the numbers on the grid and colour in all those that are multiples of 3. These numbers are members of the 3 times table.

The table normally stops at **30** in maths books, but there are also some higher numbers here that will need to be checked out.

Some important information will be shown when you have finished colouring.

- Now make up a multiplication and division statement using each of the numbers you have coloured in. For example, using the number 30, 10 × 3 = 30 and 30 ÷ 3 = 10.

- Write these down beside the number grid.

50	15	12	21	30	5
19	22	14	10	9	16
25	11	1	8	3	20
31	36	6	18	51	4
31	8	37	14	27	10
5	2	19	4	15	19
34	24	48	6	9	37
29	8	37	47	16	40
19	44	16	7	25	7
22	3	4	13	39	22
49	13	27	18	53	11
20	22	42	30	34	26
11	45	7	15	9	10
34	11	2	31	20	28

Dear Helper,

Ask your child to count through the three times table from zero to thirty, and then from thirty to zero, before they start. Ask them to count in threes beyond thirty, up to 100 if possible. Reinforce the fact that division is the inverse of multiplication, if 5 × 3 = 15 then 15 ÷ 3 = 5.

Name:

Through the maze

You will need: a coloured pencil.

- Find your way through these two number mazes from the start to the finish.

You are only allowed to move one square at a time, horizontally, vertically or diagonally.

- In the first maze follow the trail of the multiples of 3 from the start to the finish.

- Travel through the second maze from 4 to 48 using multiples of 4.

Multiples of 3:

Start

2	24	7	11	40	52	19	23	38	35	14	5	13	49	50
4	13	9	30	28	44	53	5	7	13	5	28	16	14	7
31	17	10	12	17	10	20	13	16	1	1	2	4	10	41
35	5	7	44	27	6	3	21	15	52	50	11	20	22	8
55	17	25	58	37	34	38	14	18	47	40	32	31	28	61

Finish

Multiples of 4:   **Start**

21	5	30	10	18	16	29	5	29	49	37	42	50	59	26
41	47	8	20	36	57	46	66	58	21	10	6	23	38	43
61	23	32	30	9	6	19	9	13	34	55	11	51	25	7
19	69	22	56	17	7	48	68	52	22	13	63	40	65	27
31	18	11	39	72	4	25	53	14	12	71	24	15	20	8
67	26	33	70	45	62	35	31	15	14	60	27	54	17	44

Finish

Dear Helper,

Revise both the 3 and 4 times tables with your child before they start the task. Remind them that some of the numbers they need will be higher than 10 × 3 = 30 and 10 × 4 = 40. Check that they understand the words 'horizontally', 'vertically' and 'diagonally' and that they are moving only one square at a time.

Name:

Magic squares

You will need: a pencil and either 2cm or 1cm squared paper.

You are going to investigate magic squares. They were first devised by the Chinese thousands of years ago. The squares are magic because each line, horizontal, vertical or diagonal, always add up to the same total. In the example shown each line totals 15.

- Find the totals in each of the other squares and use them to fill in the missing numbers. Some of the later squares use fractions or simple decimal numbers.

Example:

8	1	6
3	5	7
4	9	2

TOTAL = 15

①
2	9	
6		8

TOTAL = 15

②
14	11	5	
		8	
12			3
7	2		9

③
0.6		0.2
	0.5	
		0.4

④
$\frac{1}{4}$		$\frac{1}{2}$
	$\frac{5}{8}$	
		1

Dear Helper,

As an extension, ask your child to make up their own magic squares using numbers of their choice in a 3 × 3 grid. If they take a successfully completed number square and add 5 to each of the numbers, is it still a magic square? What happens if they double each of the numbers?

In the family

You will need: a pair of scissors, another piece of paper and some glue.

This task will help you understand that from one multiplication statement you can automatically make another multiplication fact and two division facts.

For example, if 5 × 3 = 15, then 3 × 5 = 15, 15 ÷ 3 = 5 and 15 ÷ 5 = 3.

- Look at the facts given in the box strips below.

- Cut out the strips and arrange them into their correct multiplication and division families. There are five different families for you to find.

- Stick the families on to another piece of paper.

6 × 2 = 12	27 ÷ 9 = 3	50 ÷ 5 = 10
28 ÷ 7 = 4	9 × 5 = 45	28 ÷ 4 = 7
2 × 6 = 12	9 × 3 = 27	45 ÷ 5 = 9
45 ÷ 9 = 5	12 ÷ 2 = 6	4 × 7 = 28
5 × 10 = 50	50 ÷ 10 = 5	27 ÷ 3 = 9
5 × 9 = 45	7 × 4 = 28	3 × 9 = 27
12 ÷ 6 = 2	10 × 5 = 50	

Dear Helper,

Through discussion, make sure that your child understands the concept that multiplication is the inverse of division and vice versa. This can be a useful tool later when checking calculations. Check the strips are correctly arranged first before they are permanently stuck into position. Your child may need to refer to a times table chart to check their solutions.

Number trios

- Look carefully at the numbers inside the boxes.

- Choose a trio of them to help you make up some multiplication and division facts. For example, if you chose 4, 3 and 12, you could use them to write 4 × 3 = 12, 3 × 4 = 12, 12 ÷ 3 = 4 and 12 ÷ 4 = 3.

- Write the facts in the space provided on the sheet.

- When you have used all these trios, you can choose some number trios of your own to give you multiplication and division facts.

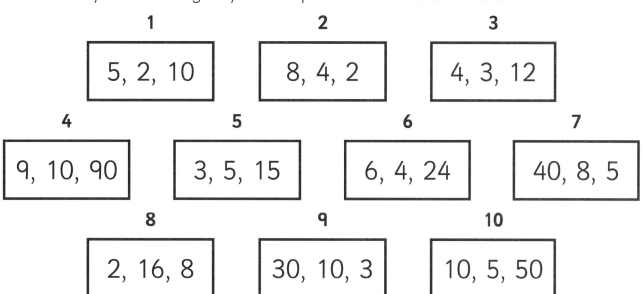

1 5, 2, 10

2 8, 4, 2

3 4, 3, 12

4 9, 10, 90

5 3, 5, 15

6 6, 4, 24

7 40, 8, 5

8 2, 16, 8

9 30, 10, 3

10 10, 5, 50

Dear Helper,

The numbers used in the boxes come from the times tables that your child has used so far. They are 2 times, 3 times, 4 times, 5 times and 10 times. Revise these tables before your child starts the activity. Remind them that each number trio should make four facts altogether, two involving multiplication and two involving division.

Name:

Double cross

You will need: a coloured pencil.

This activity looks at quick ways of multiplying numbers by using doubling and halving.

1 Multiply the numbers given below by 4. Do this by doubling them and then doubling them again. For example, 28 × 4: double 28 = 56 and double 56 = 112. Draw a line with the coloured pencil from the number to the correct answer in the box.

a 13 × 4	76
b 19 × 4	168
c 37 × 4	104
d 26 × 4	52
e 42 × 4	148

2 Multiply these numbers by 20. Do this by multiplying by 10 and then doubling the answer. For example, 14 × 20: 14 × 10 = 140, 140 × 2 = 280. Again, draw a line from the number to the correct answer in the box.

a 12 × 20	480
b 32 × 20	940
c 24 × 20	640
d 47 × 20	1 100
e 55 × 20	240

3 With this group of numbers you are going to multiply by 5. Do this by multiplying by 10 and then halving the answer. As before, link the number to the answer with a line.

a 26 × 5	215
b 57 × 5	390
c 78 × 5	180
d 43 × 5	130
e 36 × 5	285

Dear Helper,

Remind your child that one way to double and halve numbers is the 'partition method' which they may have used in the homework sheet, 'Double trouble'. Discuss other strategies they might use to find the answers. Can your child suggest other quick ways of multiplying numbers by either doubling or halving? For example, work out the 8 times table by doubling facts from the 4 times table or finding quarters by halving halves.

Name:

Top ten

When a whole number is multiplied by 10, move the digits one place to the left leaving a zero in the units column.
For example: $12 \times 10 = 120$ and $367 \times 10 = 3\,670$.

When a whole number *ending in zero* is divided by 10, move the digits one place to the right which means there is no zero in the answer. For example: $90 \div 10 = 9$ and $450 \div 10 = 45$.

- Use the function machines given below to help you remember the quick ways to multiply and divide whole numbers by 10.

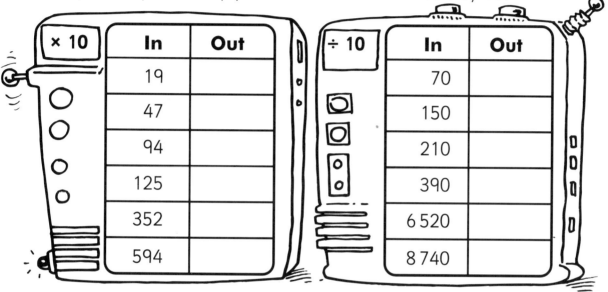

× 10	In	Out
	19	
	47	
	94	
	125	
	352	
	594	

÷ 10	In	Out
	70	
	150	
	210	
	390	
	6520	
	8740	

- Complete these statements by filling in the missing numbers.

1 ☐ $\times 10 = 230$

2 $72 \times$ ☐ $= 720$

3 ☐ $\times 10 = 6540$

4 $715 \times 10 =$ ☐

5 Make 217 ten times bigger.

☐

6 $120 \div$ ☐ $= 12$

7 ☐ $\div 10 = 230$

8 $5\,070 \div$ ☐ $= 507$

9 $9\,430 \div 10 =$ ☐

10 Find one tenth of 7 590.

☐

Dear Helper,

Revise the rules given above with your child before they start the task. Encourage them to work through the questions independently, but check over the answers with them when they have finished. Make up some questions for them to answer orally in which words are used in place of the × sign. For example, *Make 37 ten times bigger* or *What is one tenth of 3 200?*

PHOTOCOPIABLE

On the grid

This is a method to multiply TU by U, using a grid.

Remember to work out an approximate answer first.
Look carefully at the example and make sure you understand
the steps that have to be followed.
Example: **68 × 4**

Approximate answer **70 × 4 = 280**

68 × 4:	×	60	8	
	4	240	32	= 272

- Now work through these questions. Most of them are TU × U,
  but there are some to try at the end involving HTU × U.

- Work these out using the same grid method.

1 45 × 3

2 79 × 3

3 59 × 2

4 32 × 4

5 76 × 5

6 168 × 3

7 45 × 3

8 597 × 5

Dear Helper,

Allow your child to work through this practice activity on their own and then ask them to explain the
process they have used, and what they have done in each step. Check the answers with them. Discuss the
importance of finding an approximate answer first. Were they able to apply the same method successfully
when using three-digit numbers, HTU × U?

Left overs

Sometimes, when we divide whole numbers, the answer does not work out exactly.

What is left over is called a **remainder** and is always written as a whole number.

For example, 38 ÷ 4 = 9 remainder 2.

- Find the remainders in these division questions.

- In the code at the bottom of the sheet each remainder represents a letter. When you have found them all, they should spell out an important message.

1 4) 15   **2** 10) 24   **3** 3) 23

4 5) 16   **5** 7) 27   **6** 6) 23

7 10) 25   **8** 8) 39   **9** 10) 68

10 7) 33   **11** 4) 34   **12** 10) 69

- Plot the letters in order.

1 = y	5 = o	9 = k
2 = r	6 = g	10 = t
3 = v	7 = d	11 = a
4 = e	8 = w	12 = b

Message:

___ ___ ___ ___ / ___ ___ ___ ___ / ___ ___ ___ ___

Dear Helper,

Please make sure your child is showing the remainder clearly in their answer and that it is always written as a whole number. Talk about how we can check the answer to a division that includes a remainder, by using an inverse operation. For example, 53 ÷ 5 = 10 remainder 3, so 10 × 5 + 3 = 53. Ask your child to use this method to check that their answers are correct.

Big spender

You will need: a collection of loose change including the following coins:

- Work out the answers to these 'real life' money questions. Try to find at least two different ways of showing that amount, using the coins you have.

For example, if the answer was 17p, you could use 10p, 5p and 2p or you could use 5p, 5p, 5p and 2p.

1 David saves 83p one week and 45p the next. How much has he saved?

_____

2 Work out the cost of six pears, if they cost 11p each.

_____

3 A shopkeeper makes a mistake. The price tag says 99p instead of £1.49. What is the difference between the two amounts?

_____

4 Sunil has £3.00. He buys a book for £2.15. How much change does he receive?

_____

5 Nina spends 53p, £1.05 and 89p in the gift shop. How much did she spend altogether?

_____

6 Three sisters are given £2.46 to share. How much do they each get if the money is shared equally?

_____

7 If Ahmed bought a mug costing 60p, a book costing £1.20 and a card costing 80p, how much change would he have from £3.50?

_____

8 Work out the total cost of two posters at £1.75 each and three small toys at 28p each.

_____

Dear Helper,

It will help the activity to run successfully if your child is provided with five or six of each of the coins needed. Discuss the operations they chose and what strategies they used to find the answers. Can they suggest ways in which the answers might be checked, perhaps by using an equivalent calculation?

Name:

Spot the fraction

You will need: coloured pencils, squared paper, scissors and glue.

- Draw diagrams to show the following fractions, using the squared paper to help you draw the shapes accurately. Squares and rectangles will be the easiest shapes to use.

1	a half	**5**	three-eighths
2	a quarter	**6**	seven-eighths
3	three-quarters	**7**	a tenth
4	an eighth	**8**	nine-tenths

Shade in the fraction you have been given with a coloured pencil and leave the rest of the shape blank.

Cut out the shapes and stick them on to the back of this sheet.

- Here are eight counters. Colour two-eighths green, one eighth red and three-eighths yellow.

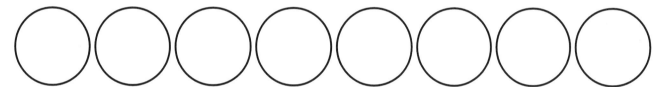

What fraction has not been coloured?

_____

- Draw ten balloons. Colour some orange, some purple and some blue.

What fractions of the balloons are coloured **1** orange **2** purple **3** blue?

1 _____ **2** _____ **3** _____

Dear Helper,

Discuss the concept of a fraction with your child before they start. A fraction is a part of a whole one. Explain that 'half' means 'one equal part out of two', a 'quarter' means 'one equal part out of four', and so on. Encourage the use of squared paper so that accurate shapes are drawn for the fractions. Check through your child's drawings and talk about how fractions go together to make a whole one.

NUMBERS AND THE NUMBER SYSTEM

FRACTIONS

Fair shares

You will need: coloured pencils, squared paper, scissors and glue.

- Draw diagrams on the squared paper to help you with these fraction questions.

- Cut out each diagram and stick it alongside the question to show how you worked out the answer.

1 Tom has 10 apples. He gives half of them to Ann. How many does Ann have?

2 Ravinder has 12 biscuits. He eats a quarter of them. How many does he have left?

3 There are 16 pears on a tree. Three-quarters of them fall off. How many are left?

4 Sanjay had £1.00 but he loses 40p. How much does he have left?

5 Clare has 24 sweets but gives three-eighths of them to Sophie. How many sweets does Sophie have?

6 There are 20 books. Mary carries one quarter of them and David carries one tenth. How many books are they carrying altogether? How many books are left?

Dear Helper,

The purpose of drawing the diagrams is to help your child see the problem in visual form. Help your child with the first question and then see if they can do the others on their own. Encourage them to use squares and rectangles to record the information, rather than complicated shapes. Remind them that fractions are closely linked to division; for example, finding half of something means dividing by 2, finding a quarter means dividing by 4, finding an eighth means dividing by 8, and so on.

Hidden fractions

- What fraction of the larger shape is the smaller one? _____

- What fraction of the larger shape is the smaller one? _____

- What fraction of the larger shape is the smaller one? _____

- What fraction of the larger shape is the smaller one? _____

- Using an arrow and the letter to indicate its position, place each of these fractions on the number line:

  A a half   **E** three-eighths

  B a quarter   **F** five-eighths

  C three-quarters **G** seven-eighths

  D one eighth   **H** four-eighths

  0 _____ **1**

Which two letters are in the same place? _____
Why? _____

Put the eight fractions into the correct order, starting with the smallest.

_____

Dear Helper,

The first part of this activity should help your child appreciate the relationship between different fractions. Some children may need to cut out the shapes and place the first on top of the second part to work out the answer. The second part of the activity will help your child begin to understand how to order fractions, that is arrange them according to their size.

Name: _____

Quick check

You will need: some coloured pencils.

Many different words are used in these questions, but they all mean that you have to carry out a subtraction.

- With a coloured pencil ring the words or phrases that tell you to take away. Then work out what the answers are.

1 Decrease 94 by 37. _____

2 From 105 subtract 68. _____

3 What is the difference between 132 and 96? _____

4 How many less than 256 is 88? _____

5 Reduce 173 by 75. _____

6 What is left if you take 56 away from 215? _____

- Write down below how you would check that your answers are correct by using an addition. Show your working out clearly.

1	2
3	4
5	6

Dear Helper,

The purpose of the number questions is to strengthen your child's knowledge of the vocabulary associated with subtraction. Please make sure your child is able to read the questions before they start. Once the answers have been calculated they need to be checked by adding. This is called the 'inverse operation'. Ensure your child can set down these checking sums correctly, working vertically down the page.

Number workout

Puzzle out the missing numbers in these addition and subtraction questions using any of the strategies or methods that you have been taught.

Remember you can check the answer to an addition by subtracting, and the answer to a subtraction by adding.

1
```
    4 5
+  □7
─────────
  1 2 2
```

2
```
    9 6
+  3□
─────────
    6 4
```

3
```
    8 4
−  3 9
─────────
    4□
```

4
```
  □2 5
+   7 8
─────────
    2
```

5
```
  4 6 7
−   3 1
─────────
  4 3□
```

6
```
  2□9
− 1 2 7
─────────
  1 5 2
```

7
```
  □9 3
+ 2 5 4
─────────
  9 4 7
```

8
```
  5 3 6
− 2□3
─────────
      6
```

Dear Helper,

Your child has been taught a number of different mental strategies and informal written methods for adding and subtracting numbers. Encourage them to use any method, or combination of methods, that will enable them to get the correct answer. Again, encourage them to self-check their solutions by using the inverse method, subtraction in adding sums and addition in subtractions.

PHOTOCOPIABLE

57

Just in time

1   Draw the minute and hour hands on these clock-faces to show the times given underneath. Remember that the minute hand should be longer than the hour hand.

3 o'clock   **10 past 5**   **quarter to 7**   **half past 10**

2   These are the faces of digital clocks. Write numbers in the boxes to show the times given underneath, as they would appear on a 12-hour clock.

:	:	:	:

20 to 9 $\frac{1}{4}$ **past 6**   **4 o'clock**   **ten to 11**

3   Look carefully at the position of the hands on these clock-faces and then write down the times underneath.

_____   _____   _____   _____

4   Look carefully at the faces of these digital clocks and then write the times in words underneath.

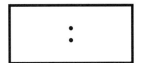   

3:25   9:05   2:15   11:55

_____   _____   _____   _____

Dear Helper,

In this activity your child will practise writing and reading times using both an analogue and a 12-hour digital clock-face. Check your child's ability to read both types of clock-face successfully, using different clocks around the home, before they start. Work with your child, helping with things such as the position of the clock hands and important spellings like half, quarter, minute, twenty, thirty, forty and fifty.

Name:

Busy day

You are going to keep a time log to show how you spend a typical school day.

- Fill in the three columns in the log below by writing down the activity you are doing, what time you started it, and then, as accurately as you can, how long it took you to complete.

The first column has been started for you.

What I did	When I started	When I finished	How long it took
1 Got out of bed.			
2 Had breakfast.			
3 Travelled to school.			

Dear Helper,

Your child has been asked to complete a simple personal time log for a day. Explain to them that only main events in the day should be included and that they will not be able to put down everything. Help may be needed with the way in which times are written, especially the use of a.m. and p.m., and when they are working out how long different activities took.

Name:

Finding out

The chart shows the results some children gained in their mental arithmetic and spelling tests. The mental arithmetic test was out of 20 and the spelling test was out of 30.

There are three spare columns for you to use as you wish. They should help you to organise the data so that you can answer the questions at the bottom of the chart. You could use them, for example, to give the children's total marks.

Name	Boy/Girl?	Mental arithmetic	Spelling			
Rachel	Girl	15	25			
Sunil	Boy	19	28			
Aslam	Boy	12	29			
Paula	Girl	10	13			
Nina	Girl	17	19			
Kate	Girl	16	27			
Ian	Boy	9	21			
Adam	Boy	14	12			
Sita	Girl	18	30			
Claire	Girl	15	14			

Answer these questions a separate sheet of paper.

1 Who scored the highest mark in mental arithmetic?
2 Who scored the highest mark in spelling?
3 Who is the best all-round pupil?
4 Who is the worst all-round pupil?
5 How many children scored more than half marks in the mental arithmetic test?
6 How many children scored more than half marks in spelling?
7 How many girls scored more than half marks in spelling?
8 How many boys scored more than half marks in mental arithmetic?
9 Most boys did better in arithmetic than spelling. True or false?
10 Most girls did better in spelling than in arithmetic. True or false?

Dear Helper,

This activity gives your child the opportunity to interpret and analyse a set of data or information. Discuss with them what the information shows and how they can use the spare columns to help organise the data so that they can answer the questions more easily. Please make sure that they understand the meaning of the phrases 'all-round' and 'half marks'.

Happy birthday

Some children at Westway School have been collecting information about when their friend's birthdays are. They have spoken to 40 children and made a tally chart to show which months their friends were born in.

The tally chart they made is shown along side.

Find the total number of children born in each month and then answer the questions. You may need to work on a separate sheet of paper.

Name	Tally	Total
January	|||	
February	⊬⊦⊦ ||	
March	|||	
April	⊬⊦⊦ ||||	
May	||	
June		
July	||	
August	||||	
September	⊬⊦⊦	
October	||	
November			
December	||	

1   How many children were born in September? _____

2   How many children were born in December? _____

3   In which month were most children born? _____

4   In which months were only two children born? _____

5   In which month were no children born? _____

6   How many children were born during the 'summer' months of June, July and August? _____

7   How many children were born during the 'winter' months of November, December, January and February?

_____

8   Make up your own question that could be answered by using the chart. You could use your own birthday month.

Dear Helper,

The tallying system used here is a quick and reliable way of collecting information for data-handling activities. It will become particularly useful when your child is dealing with larger numbers. Ask them to explain how the system works, what the symbols mean, and what they feel the advantages of tallying are. Discuss the questions with them, especially number 8.

Name:

Get into shape

Lauren and Asif have been very helpful to their teacher. They have counted up all the plastic and wooden 3-D shapes in the maths resource cupboard so that the teacher will know how many new shapes to order as lots were missing.

- Make a pictogram on the back of this sheet to show how many different shapes they counted. Use the information below.

- Now explain how this information helped the teacher with her order.

3-D maths shape	Total counted
1 Cube	**15**
2 Cuboid	**12**
3 Tetrahedron	**10**
4 Egyptian pyramid	**8**
5 Triangular prism	**1**
6 Sphere	**5**
7 Cone	**9**
8 Cylinder	**16**

Dear Helper,

Your child will already have made a pictogram in class. In a pictogram, picture symbols are used to represent the numbers involved. Items can be shown on a one-to-one basis, one drawing for every item, but with larger numbers a symbol can represent two, five, ten or twenty units. Help your child to give the pictogram a title and to label both the horizontal and vertical axes carefully. They should also write down what each picture symbol stands for.

Sporty types

- You are going to make a pictogram of your friends' favourite sporting games.

- Survey only your friends at school and members of your own family at home to find out which sports they enjoy best. *Do not ask anyone else.*

- Show these sports on the pictogram by drawing a picture of some of the equipment people use when they are playing it, for example, a football for football, a tennis racket for tennis and a club for golf and so on.

- Remember to give your pictogram a title, to label the axes to indicate what information you are showing, and to explain what each drawing represents.

Use the squared paper below for your pictogram.

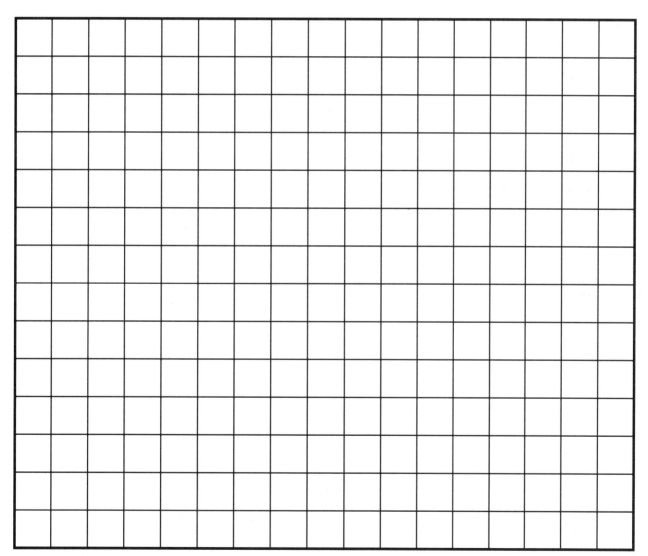

Dear Helper,

This time your child has to gather their own information to make the pictogram. Help them to decide what each diagram will represent on the pictogram, how they will label the axes and what title they will give the piece of work. In class, the children will be doing further follow-up work using the pictogram.

Comparison sentences

You need to know the meaning of these symbols to do this homework.

 **<** means 'less than'   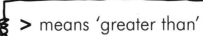 **>** means 'greater than'   **=** means 'the same as'

Here are some number sentences:

Jack and Jill are twins.
They are 9 years old.

Raj and Sita are brother and sister. Raj is 14 and Sita is 5.

 Raj's age > Sita's age, because 14 > 5.

 Jack's age = Jill's age, because 9 = 9.

 Jill's age < Raj's age, because 9 < 14.

- Write some sentences involving ages or measurements like these, using the symbols < > = in the space below.

Dear Helper,

This activity will help your child to learn how to use the symbols < > and = and to manipulate numbers. Please discuss the homework and suggest numbers they can compare. Look for ideas around the home. You may use measurements comparing the weights, quantities or volume of items. Ask your child to explain the comparisons verbally, before trying to record them using the symbols.

ORDERING AND ROUNDING   NUMBERS AND THE NUMBER SYSTEM

PHOTOCOPIABLE

Name:

Where is it coldest?

You will need: some information about the temperatures of places around the world. Get this from newspapers, TV, radio, Teletext or the Internet.

- List information about the temperature in ten different places.

- Which place do you think is the coldest?

- Which is the warmest?

- Can you write them in order, starting with the coldest?

Dear Helper,

Please help your child to find information about the temperature of different places around the world. It would be particularly helpful to find some very cold places with negative temperatures, such as –15°C, as the children need to be able to recognise negative numbers and place them in order. For example, –15°C is colder than –4°C.

Name:

Making sums

- How many different sums can you make using
  the numbers 25, 37 and the sum of the two.

For example: 25 + 37 = 62,   62 – 37 = 25,   62 = 37 + 25.

There are eight possible arrangements!

Now try these other pairs of numbers:

- Write down all the different sums you can make
  from these pairs of numbers and their sum.

Dear Helper,

This activity will help your child to understand the relationship between addition and subtraction and give them a method of checking calculations by using the inverse, or opposite, operation. For example, check an addition by subtracting one of the numbers from the total. You can suggest further pairs of numbers to extend the activity and you may wish to introduce three-digit numbers if your child is confident. If necessary, explain the activity by using simple addition of single digits: 5 + 2 = 7, 2 + 5 = 7, 7 – 2 = 5 and so on.

Name:

Double up

You will need: a dice.
You can play this game on your own or with a partner.

- Take it in turns to roll the dice twice to find a number. The first roll will give you the tens digit and the second roll will give you the units digit.

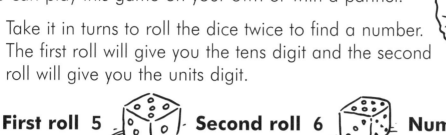

First roll 5   **Second roll** 6   **Number** 56

- Write down the number and then see if you can 'double up'.

- If you are playing on your own, make ten numbers, double them and then ask an adult to check them for you.

- If you are playing with a partner you can check each other's work and see who gets the most right!

Start number	Double it	Start number	Double it

Dear Helper,

We have been doubling numbers up to 50 in class. This game will help your child practise 'doubling' which is a strategy that can be used to work out a whole range of calculations. Encourage your child to use their creativity to work out the doubles and to talk about the strategies used. You may like to challenge them by suggesting numbers to 'double'.

Name:

Work it out!

Here are some sums.

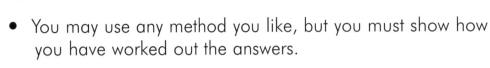

- You may use any method you like, but you must show how you have worked out the answers.

- See if you can find the easiest method for each question!

1	27 + 39 + 1

2	199 + 29

3	350 + 150

4	34 + 56 + 40

5	60 + 70 + 40

6	78 + 35

7	64 + 56

8	25 + 25 + 30

9	75 + 36 + 25

10	79 + 99

Dear Helper,

Encourage your child to look at each sum and think about the easiest way to calculate that particular addition. Help them to think about pairs of numbers that make 10 and 100, about doubles, and when it is best to round up or down and then adjust the answer. It is important that the child records their calculations, but this does not have to be in a formal way.

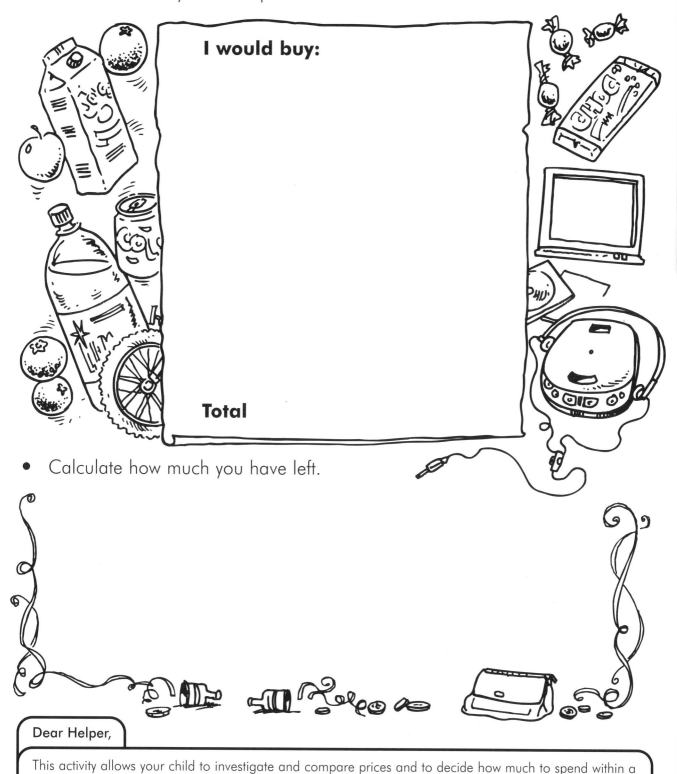

Spending your birthday money

Imagine that you have been given £50 as a birthday present.

- Decide how you would like to spend your money. You can look in catalogues, the newspaper or in local shops. You have to choose at least three items and you must spend at least £45.

I would buy:

Total

- Calculate how much you have left.

Dear Helper,

This activity allows your child to investigate and compare prices and to decide how much to spend within a limited budget. This will involve adding up money. Help them to look for the cost of items and make decisions about what can be purchased. You may suggest that they make two lists, one with a single large item and several smaller ones, another with as many items as possible for the money!

Name:

Units first

Here is a way of adding HTUs:

```
  H T U
  2 4 7
+ 3 3 8
    1 5    adding the units first
    7 0    adding the tens next
  5 0 0    then adding the hundreds
  5 8 5    and then adding the three numbers to give the answer.
```

● Complete these sums in the same way, remembering to add the units first.

1
```
    4 2 4
  + 4 4 5
```

2
```
    2 7 4
  + 6 2 3
```

3
```
    5 2 5
  + 3 6 5
```

4
```
    6 4 7
  + 2 5 2
```

5
```
    4 8 2
  + 4 1 8
```

6
```
    6 2 6
  + 3 6 6
```

7
```
    5 6 7
  + 2 3 3
```

8
```
    5 3 9
  + 2 5 5
```

Dear Helper,

This activity introduces the standard written method for column addition with which you will be familiar. However, here it is written in an extended form, rather than using 'carrying'. Encourage your child to use this version of the method for now and to write the sum out in full, as given in the example. This will ensure that your child really understands the stages before progressing to the more familiar shortened version.

Exchange it!

536 =	500 + 30 + 6	'536 is five hundreds, thirty and six.'
– 48	– _____ 40 + 8	'minus 48, which is forty and eight'
=	500 + 20 + 16	We can exchange one 10 from the
	– 40 + 8	thirty, leaving twenty, and add it to the six to make sixteen.
=	400 + 120 + 16	We can exchange one hundred from the
	– 40 + 8	five hundred, leaving four hundred, and add it to the twenty to make 120.
= 488	400 + 80 + 8	When we subtract each column the result is 488.

- Using the method above, and setting your work out in this way, calculate the following subtractions. Use more paper if you need more room for your working out. There is no need to write the number words.

1

```
   4 9 6
 – 1 7 7
 _____
```

2

```
   9 3 8
 –   8 9
 _____
```

3

```
   6 2 9
 – 4 3 9
 _____
```

4

```
   3 8 7
 – 1 8 9
 _____
```

5

```
   8 3 6
 –   6 8
 _____
```

6

```
   2 8 8
 – 1 9 3
 _____
```

Dear Helper,

This exercise is for your child to practise using the 'decomposition' method for column subtraction. Please encourage your child to set the sum out in the way illustrated as this will aid understanding. Some further examples that you may encourage your child to try are: 736 – 157, 492 – 277, 583 – 196.

PHOTOCOPIABLE

Name:

What measurements do we use?

You have been learning about different units of measurement such as: metres, centimetres, grams, kilograms and litres.

- Look around your home and see what items you can find with measurements marked on them.

- Make a list of the different items and their measurements under these three headings.

Mass (e.g., sugar 1kg)	Length (e.g., kitchen roll 5m)	Capacity (e.g., fruit squash $1\frac{1}{2}$ l)

Dear Helper,

This activity will help your child recognise different units of measurement in context. Help your child to make comparisons between different items and discuss the units marked on them and how they are recorded. For example, a bag of sugar may be labelled with 1 kilogram (1kg), but a packet of butter is labelled in grams (e.g., 450g).

Name:

How long is it?

You will need: someone to work with,
a measuring tape and a ruler.

- Try to estimate the length of items around the home.
  For example, how long is the window?

- Write down your estimate and then check the
  measurement with a ruler or measuring tape.

- Record your estimates and your measurements.

Item	Estimate	Measurement
Window	$1\frac{1}{2}$ m	130cm (or 1.3m)

Dear Helper,

Please work with your child, encouraging estimation of the lengths and then checking by measuring.
Talk about the different units, cm and m, making sure that your child recognises that there are 100cm
in 1 metre. Encourage them to guess in fractions of a metre, such as ½ (50cm), ¼ (25cm) or ³⁄₁₀ (30cm).
Set challenges, for example: *Can you guess the length of the table to within 2cm of its length?* Let your
child challenge you!

Name:

Measuring up!

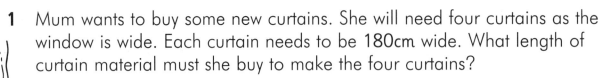

- Work out the answers to these problems, showing your calculations. Remember to think about the units and how you record them.

1 Mum wants to buy some new curtains. She will need four curtains as the window is wide. Each curtain needs to be **180cm** wide. What length of curtain material must she buy to make the four curtains?

_____

2 The parking deck on the car ferry is **50m** long. The deck is wide enough for two cars to park side by side. Each car must have a space **5m** long. How many cars can park on the deck?

 _____

3 One bottle of fizzy drink holds **2l**, each mug holds **200ml**. How many bottles of drink do we need to fill **16** mugs?

_____

2 litres 2 litres

4 Dad's suitcase weighs **15kg**, Mum's weighs **19kg** and Ben's weighs **12kg**. How much does their luggage weigh altogether?

_____

5 Shazia wants to buy a mixed bag of sweets for her party. She buys **100g** of humbugs, **150g** of toffees, **225g** of fruit drops and **200g** of chocolate limes. How much did her mixed bag of sweets weigh?

_____

Dear Helper,

This is an exercise to help your child work out 'real life' problems using measures. If they need help, encourage them to think about the information that is given in the question and decide which are the important parts. Help your child to identify what calculations need to be done and make sure that they understand the units in the question.

Name:

Draw and count

You will need: a pencil and some objects to draw around.

- Find at least four small objects that you can draw around.

- Carefully draw around each object on to this squared paper.

- Count up the squares (cm²) to find an approximate area for each shape.

Remember that if a part of a square is more than half a square you count it as 1, but if it is less than half a square you ignore it.

Dear Helper,

This is an introduction to finding areas. Help your child to select suitable objects to draw around (not too complicated!) Ask them to number the whole squares as they are counted, then to look carefully at the part squares. Help them to decide whether the squares are more or less than a half.

Make a picture

• Draw a simple picture on this grid and then use co-ordinates to give instructions to your Helper, so that they can copy your picture. For example, you may draw the outline of a house:

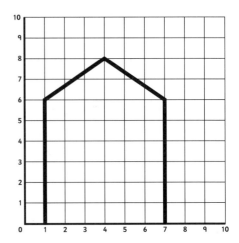

**Start at (1,0), draw a line to (1,6),
draw another line to (4,8),
then one to (7,6),
and finally to (7,0).**

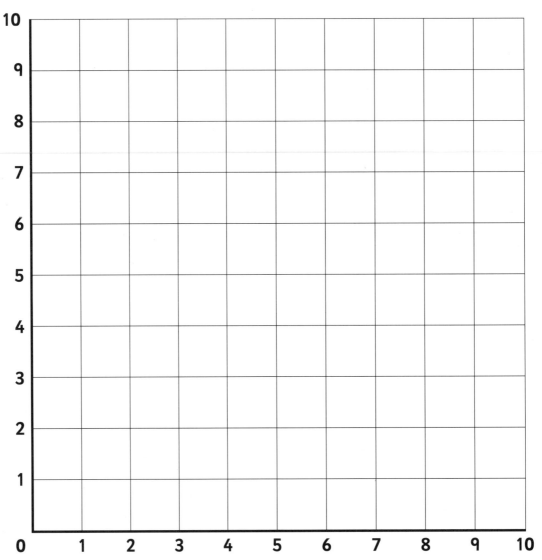

Dear Helper,

This activity will help your child to learn about co-ordinates. Let them practise with you by drawing simple shapes on squared paper and working out the co-ordinates of different points on the shape. Then see if they can draw a shape following your instruction. Finally challenge them to give correct instructions to enable you to draw the picture that they have designed. If necessary, remind your child to count along and then up.

Name:

Which direction?

St John's Church		Page Park		Peter's House
Cheapo Supermarket		Uptown School		Post Office
Ahmed's House		Town Centre		Cinema

1 Who lives to the NE of Uptown School? _____

2 What is due north of Uptown School? _____

3 What is to the west of Page Park? _____

4 What is NW of the town centre? _____

5 Who lives due south of the supermarket? _____

6 In which direction does Ahmed have to walk to get to the cinema? _____

7 Which direction does Peter have to walk in to get to Ahmed's house? _____

8 What building is SE of Page Park? _____

Dear Helper,

This is an exercise to help your child practise finding information using simple compass directions. If they find it difficult, help to identify where each place is in relation to the school. First find the four basic directions of N, S, E and W and then help the child to identify NW, NE, SW and SE.

Shape collection

Here are some common shapes:

cuboid,   cone,   sphere,   square,   cube,   circle,   cylinder,   rectangle

- Write down some general statements that describe them.

   For example: 'Cuboid: a 3-D shape, six rectangular faces, opposite pairs of faces are the same as each other.'

Cone

Sphere

Square

Cube

Circle

Cylinder

Rectangle

- For each shape find some items around the home that satisfy these statements. **For example:** Cuboid: cereal box, chest of drawers.

Dear Helper,

Make sure that your child can describe each of the shapes, before looking for items to match the statements. This activity will help your child to identify the shapes with different dimensions in context, for example, recognising that a shallow cylindrical tin (such as a tin of pins) is a cylinder which satisfies the same general statements as a tall cylinder (such as a carton of salt).

Name: _____

Find the rule

Here is a sequence of numbers:

1, 5, 9, 13, 17 The **rule** for this sequence is **add 4**.

- Play the game 'Find the rule' with your Helper.

- Take it in turns to write down a sequence and see if the other person can work out the 'rule'.

- To start you off, here are some sequences for you to think about. Can you spot the rule?

  2   4   8   16   32   Rule: _____

99 96 93 90 87   Rule: _____

80 40 20 10   5   Rule: _____

Your go…

Dear Helper,

Play this game with your child and encourage them to think about sequences with different starting points. An easy way to spot the rule is to find the differences between consecutive numbers. Initially, you could just use a rule with steps of constant size such as 'add 5', or 'subtract 10'. When your child is confident, use other rules such as halving, doubling or multiples.

Name:

Odds and evens

Here is a game to play with your Helper.

You will need: a ten-sided dice, numbered 0–9.

- Take it in turns to roll the dice to generate a two-digit number (for example, 2 then 6 will give 26).

- Predict whether the number will be odd or even **before** you roll the dice.

- If you are correct you win a point, if not the other player wins the point.

Keep a record of your points on this chart.

Player one			Player two		
Prediction	Number	Points	Prediction	Number	Points
Odd	53	1	-		0
-		1	Odd	28	0
Even	27	0	-		1
-		0	Odd	55	1
Even	86	1	-		0
-		0	Odd	45	1

Player one			Player two		
Prediction	**Number**	**Points**	**Prediction**	**Number**	**Points**
Total			**Total**		

Dear Helper

Your child has been learning about recognising odd and even numbers. Play this game with your child, encouraging them to try to predict whether the number generated will be odd or even. You can extend this game, to include numbers greater than 100, by rolling the dice three times.

Name:

Number puzzles

Here is a number code:

A = 1	B = 2	C = 3	D = 4	E = 5	F = 6	G = 7	H = 8
I = 9	J = 10	K = 11	L = 12	M = 13	N = 14	O = 15	P = 16
Q = 17	R = 18	S = 19	T = 20	U = 21	V = 22	W = 23	X = 24
Y = 25	Z = 26

- Solve these problems to find the key word.

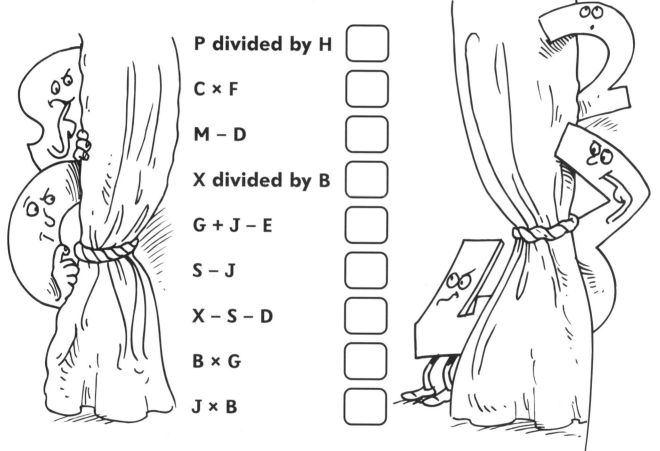

P divided by H

C × F

M − D

X divided by B

G + J − E

S − J

X − S − D

B × G

J × B

- Make up some secret messages like this of your own.

Name:

Tables bingo

- You will need to make your own bingo card.

  Fill in any ten numbers on a blank card.

  Like this:

	8		15		24	
10		18		21		40
	45		50		72	

Do not show your Helper your bingo card yet.

- Ask your Helper to give you some multiplications to do.

- If the answers are on your card, cross them out.

Dear Helper,

This activity will help your child to recall multiplication facts quickly. Say, for example, 'Five times two.' Your child crosses out the number 10 if they have it written on their card. Initially, you may choose to limit the game to one multiplication table, for example the 5 times table, in which case tell your child to put only numbers from that 'table' on their card. When you are practising a range of tables, encourage your child to think about which are the 'better' numbers to choose, for example 20 occurs in the 2, 4, 5 and 10 times tables, but 15 only occurs in the 3 and 5 times tables.

Name:

Rearrange it!

Here is a way of rearranging a multiplication to make it easier to work out:

8 × 16 = 8 × (4 × 4) = (8 × 4) × 4 = (8 × 4) × (2 × 2)

= 32 × 2 × 2 = 64 × 2 = 128

- Rearrange these multiplication sums to make them easier to calculate:

9 × 16
15 × 6
30 × 8
12 × 9
18 × 9
12 × 15

Dear Helper,

Your child has been learning about reordering multiplications. This activity will provide practice in manipulating multiplications to find the easiest method of calculation, using multiplication facts as well as doubling. If your child finds this easy, let them investigate using the method with larger numbers, such as 16 × 24.

Name:

Using the grid!

Here is a grid method for multiplication:

17 × 9

×	10	7	Total
9	90	63	153

- Do these multiplications using the grid method.

18 × 7

×			Total

15 × 8

×			Total

33 × 9

×			Total

25 × 9

×			Total

67 × 8

×			Total

49 × 7

×			Total

Dear Helper,

This method of multiplication helps to prepare your child to use the more traditional methods, particularly thinking about place value. Encourage your child to break the two-digit numbers down into tens and units before multiplying.

Name:

Approximate first!

Here is a standard method for multiplying 68 × 9:

Approximate first:
68 is nearly 70, 70 × 9 is 630.

612 is close to the approximation of 630.

- Do these multiplications using this method, remembering to approximate first and then calculate.

---

1) **59 × 7**

2) **95 × 8**

3) **73 × 6**

4) **84 × 9**

5) **66 × 5**

6) **93 × 6**

Dear Helper,

See how quickly your child can do these multiplication sums using this method. It is important to encourage them to approximate first and then to check their calculation against the approximation, to see whether the answer 'looks right'.

Check it!

Remember that it is always a good idea to use another calculation to check an answer.

Sometimes you may check using a different method, at other times you can use an **inverse** calculation.

For example, 12 × 5 = 60 can be checked by dividing 60 by 5.

- Work out each of these calculations and then check your answer, either by using a different method or by doing an inverse calculation.

	Check it!
1) 49 × 7	
2) 76 ÷ 4	
3) 28 × 8	
4) 96 ÷ 8	

Dear Helper,

Your child has been using different methods for multiplication and division and has been learning how to check their answers for accuracy. Encourage your child to think about the best way to check an answer, discussing different methods. You may like to work through some other examples with your child, pointing out the relationship between addition and subtraction, and between multiplication and division.
Try some like these: 93 – 38, 69 + 38, 45 × 9, 64 ÷ 4, 36 × 6, 120 ÷ 10.

100 MATHS HOMEWORK ACTIVITIES • YEAR 4 TERM 2

Name:

What's left?

You will need: a ten-sided dice and a multiplication square.

Play this game with your Helper.

- Take turns. First you roll the dice twice to generate a two-digit number.

- Your partner then rolls the dice once to find the divisor (the number to divide by).

- You now do the division and work out the **remainder**. This gives your point score for the turn: the bigger the remainder, the bigger the score! Keep a note of your scores on the back of this homework sheet.

- Your partner now has a turn at rolling the two-digit number.

- The winner is the person who reaches 20 points first.

   For example:

   You roll 2 and 7. Your start number is 27.

   Your partner rolls 6. You work out 27 ÷ 6 and get 4 remainder 3.

   Your score is 3 points.

Dear Helper,

Your child has been learning about remainders after division. This game will help them to work out divisions and remainders quickly. Encourage them to check the answer by using a multiplication square. See if they can recognise multiples quickly. You may like to make this easier by using a six-sided dice or by limiting the divisors to the 'easier' times tables such as 2, 5 and 10.

Name:

The school trip

- Here are some problems where you have to make sensible decisions about rounding up or down.

- Remember to show how you worked the problems out.

  There is going to be a school trip. Four classes are going. Each class must have at least one adult for every six children.

- Work out how many adults each class will need to take.

Adults needed

Class 1A has 27 children.

Class 2B has 32 children.

Class 3C has 31 children.

Class 4D has 25 children.

You have to arrange the transport.

A 33-seater coach costs £100, whereas a 49-seater coach costs £150.

- Using which size of coach would be the cheaper? _____

- What would be the least amount of money we would have to spend on coaches? _____

  Each class would like to travel on a separate coach.

- Which coaches should we book? _____

- How much will they cost? _____

- What do you think each child should have to pay to go on the trip?

  _____

- It is easiest to charge an amount to the nearest 10p. What would you charge? Explain why.

  _____

Dear Helper,

This investigation will make your child consider whether to round up or down when there is a remainder after division. Help them to think about the different possibilities for the coaches and to come up with their own idea about what the trip should cost.

100 MATHS HOMEWORK ACTIVITIES • YEAR 4 TERM 2

Name:

Fraction match

- Colour and label this fraction board as quickly as you can.

- Circle the fraction in each coach that is equivalent to the first fraction in the train. How long does it take? _____

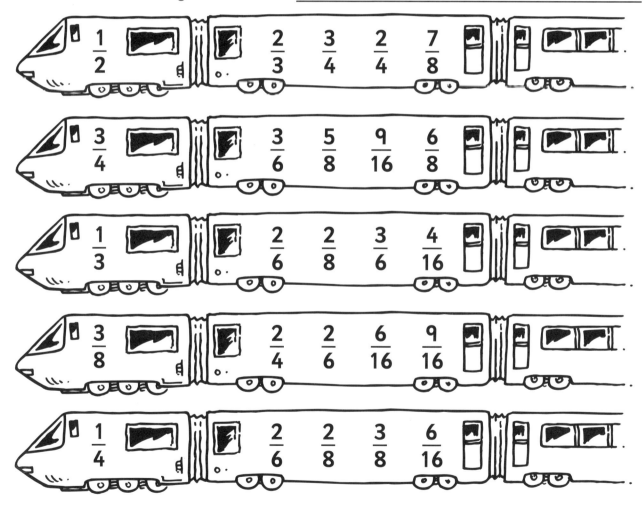

Dear Helper,

Your child has been learning about fractions and matching simple fractions with equivalents. Help them to complete the fraction board, colouring and labelling each section using ½s, ¼s, ⅛s, ¹⁄₁₆s, ⅓s, ⅙s and their multiples. Go through the various equivalencies on the fraction board and encourage your child to use the fraction board to complete the exercise.

Name the fraction!

- Colour in some sections in each square and then label the coloured fraction. Two have been done for you.

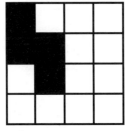

$\dfrac{4}{16}$

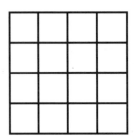

$\dfrac{1}{4}$

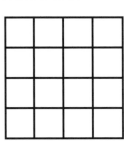

_____

_____

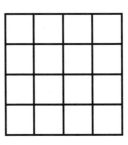

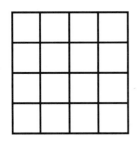

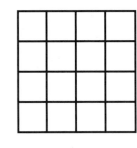

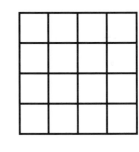

_____

_____

_____

_____

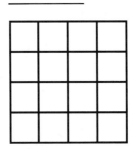

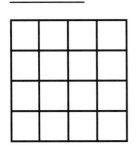

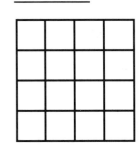

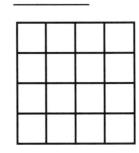

_____

_____

_____

_____

Dear Helper,

This activity will help your child to recognise and name fractions. It is important that they realise that fractions can be different shapes, and that a fraction is part of a 'whole'. Encourage your child to find interesting ways of colouring the squares to make fractions. Draw some other grids for them to try, for example 3×3 or 3×4 and talk about the fractions coloured in.

Name:

Make 1

- Label each of the fractions and then find pairs that total 1.

a
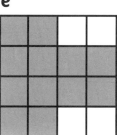

b
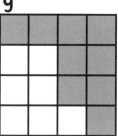

c
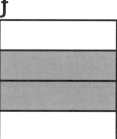

d
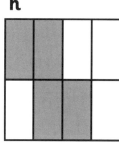

e
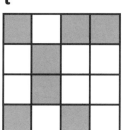

f
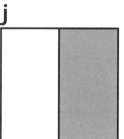

g
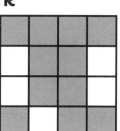

h
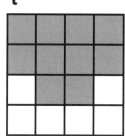

i
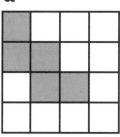

j
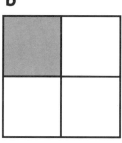

k
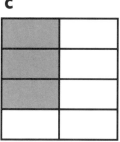

l
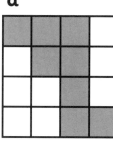

- Write your pairs of fractions making a total of 1 here:

Dear Helper,

Encourage your child to try to identify the fractions, counting the squares if necessary. The aim of the activity is to identify pairs of fractions that make 1. This will help your child to recognise equivalent fractions and to understand the concept of fractions as parts of a whole.

Name:

Using fractions

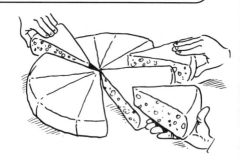

- Answer the questions below.

1 Find $\frac{1}{4}$ of 16.

_____

2 What is $\frac{1}{8}$ of 24?

_____

3 How many centimetres are there in a quarter of a metre?

_____

4 What is $\frac{1}{2}$ of 12?

_____

5 How many grams in $\frac{1}{4}$ kg?

_____

6 How many minutes in half an hour?

_____

7 Find $\frac{1}{3}$ of 15.

_____

8 What is $\frac{1}{10}$ of 100?

_____

9 What is $\frac{1}{4} \times 20$?

_____

10 What is $\frac{1}{2} \times 50$?

_____

Dear Helper,

This practice exercise will help your child to understand fractions in 'context' and to begin to relate simple fractions to division. If your child has difficulty with this exercise, talk about what each question actually means, giving some further examples if it helps. Time how quickly your child can complete this task.

Name:

Pictogram

Some children have collected information about the hobbies of their classmates.

- Can you complete the pictogram using this data?

- Think carefully about the scale you use and keep your 'pictures' simple.

Hobby or interest	Number of children who take part
Football	16
Cycling	6
Horse riding	4
Cricket	12
Playing the keyboard	7
Swimming	18
Computer games	15

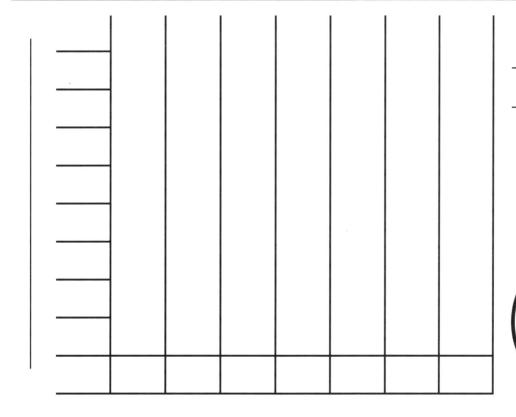

Don't forget to label the axes and give your pictogram a title.

- Think of some questions that you could ask about the pictogram.

Dear Helper,

Your child has been learning about graphs and pictograms. Help them to decide the most appropriate scale and a suitable icon (or small picture). Each 'icon' could be equal to 1 or 2 units. Talk about the information and ask questions such as: *Which is the most popular hobby?* or *How many children play a ball game?*

HANDLING DATA

INTERPRETING DATA

Which books?

- You are going to collect data from books you have at home, to use in school for some graph work.

- You will need to design a data collection sheet or a 'tally sheet' to record your data.

- Decide which categories you are going to use to collect the data and then do your own 'book count'. You may decide to use types of book or authors.

For example:

Adventure stories	ЖЖ |||
Science fiction	|||
Nature books	ЖЖ ЖЖ	
Poetry books	||

Data collection sheet

Category	Tally	Total numbers

PHOTOCOPIABLE

Dear Helper,

You can talk to your child about the size of the 'book count'. You may wish to limit the sample of books to a selected pile or you could extend it to include a whole bookcase! Help your child to decide the categories that most interest them.

Name:

Graph it!

- Use the data you collected in your 'Which books?' homework to produce some graphs of your own.

  You may need some extra paper.

- Start with a bar chart.

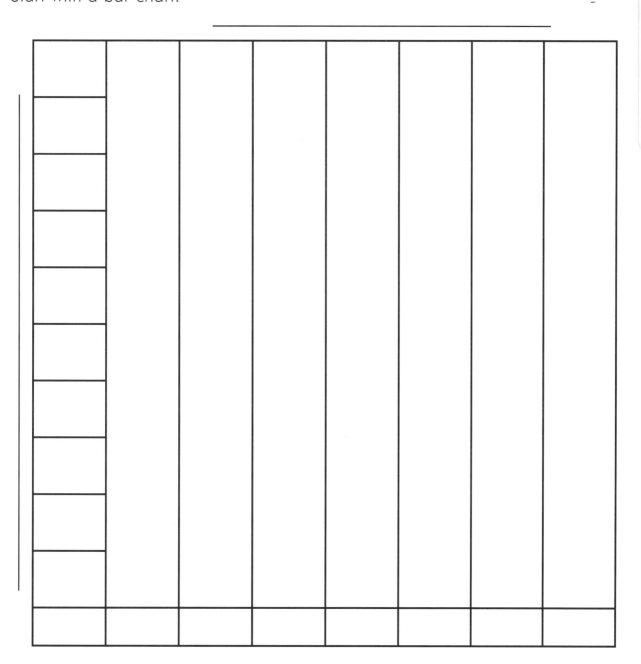

- Can you think of any other ways to use the information in a graphical form?

Dear Helper,

This activity will give your child further practice with graphs. Discuss the various graphs that can be drawn. You may suggest that they draw a pictogram with each 'book' icon representing two or more books. Encourage your child to collect data and draw some other graphs.

Name:

Ton up

You will need: a coloured pencil.

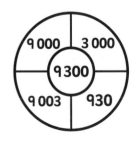

Remember that when you multiply 100 by whole numbers, the digits move two places to the left to correspond with the number of zeros. So $3 \times 100 = 300$ and $28 \times 100 = 2800$.

- Multiply the number on the arrow by 100 and draw a line with your coloured pencil to show the correct part of the target where it should land.

- Now colour in that part of the target as well.

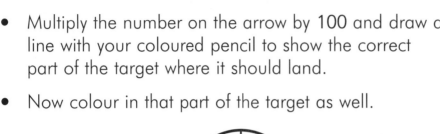

1 ➤ **5** ➤ | 15 | 50 | 25 | 500 | 51

5 ➤ **36** ➤ | 3 000 | 6 000 | 3 600 | 3 060 | 1 360

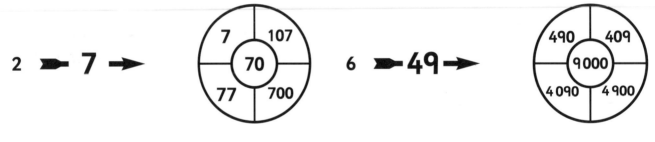

2 ➤ **7** ➤ | 7 | 107 | 70 | 77 | 700

6 ➤ **49** ➤ | 490 | 409 | 9 000 | 4 090 | 4 900

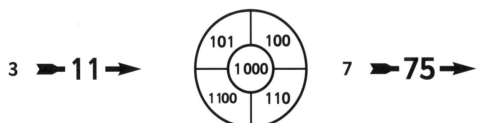

3 ➤ **11** ➤ | 101 | 100 | 1 000 | 1 100 | 110

7 ➤ **75** ➤ | 7 500 | 7 005 | 700 | 1 700 | 750

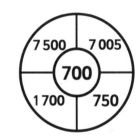

4 ➤ **14** ➤ | 1 400 | 1 040 | 141 | 1 444 | 140

8 ➤ **93** ➤ | 9 000 | 3 000 | 9 300 | 9 003 | 930

Dear Helper,

You may need to remind your child that when a whole number is multiplied by 10 the digits move one place to the left, for example: $5 \times 10 = 50$ and $23 \times 10 = 230$. Then revise the information shown at the top of the sheet about multiplying by 100. Practise some examples. Please stress to your child that these quick methods can only be used when dealing with whole numbers.

Give me a sign

1 Put the correct sign <, > or = between these pairs of numbers.

105	115

142	124

656	566

1 702	2 107

9 009	9 900

double 14	28

45	double 22

2 Put in the missing signs <, > or = in these questions.

52 + 47	100

155	200 − 51

253	146 + 98

50 ÷ 5	44 ÷ 4

double 18	9 × 4

194 − 77	121

7 × 4	9 × 3

half of 64	double 30

3 Complete these number sentences using your own numbers.

145 + 69 >

> double 27

356 − 78 =

42 × 6 <

= 126 ÷ 3

half of 74 >

Dear Helper,

Discuss what the signs show. < means 'smaller than' and > means 'bigger than', with the open part of the sign always going towards the larger number. For example: 9 > 6 and 10 < 12. Also discuss the equals sign (=) and when it is used. Encourage your child to say the statements aloud as they complete the questions. For example: 15 > 9, 'fifteen is bigger than nine' and 17 < 21, 'seventeen is smaller than twenty-one'.

Top tens

You are going to add together three two-digit multiples of ten, e.g., 50 + 30 + 60.

- Choose these numbers from the stars. Their totals are shown in the planets.

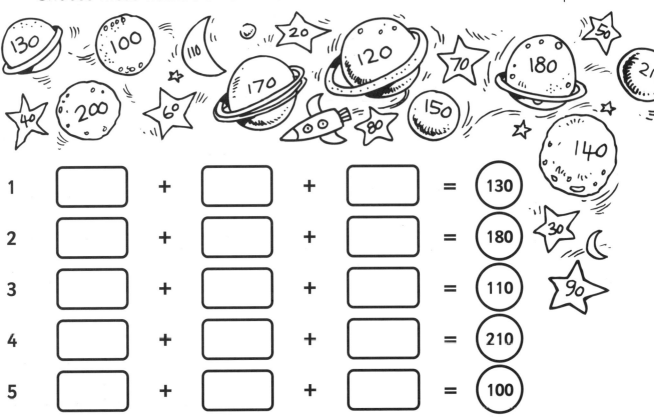

1 ☐ + ☐ + ☐ = ⬤ 130

2 ☐ + ☐ + ☐ = ⬤ 180

3 ☐ + ☐ + ☐ = ⬤ 110

4 ☐ + ☐ + ☐ = ⬤ 210

5 ☐ + ☐ + ☐ = ⬤ 100

In these questions one of the star numbers is given.

- Find the other star numbers and the total that needs to go into the planet.

6 ☐ + 30 + ☐ = ◯

7 20 + ☐ + ☐ = ◯

8 ☐ + ☐ + 50 = ◯

9 ☐ + 80 + ☐ = ◯

10 60 + ☐ + ☐ = ◯

Dear Helper,

Work with your child to help them add the three two-digit multiples of 10 in their head as a mental calculation. Both of you should work out the answers separately and then discuss findings. Talk about checking the answers by using the 'inverse operation', i.e., the opposite. For example, if 70 + 40 + 20 = 130, then 130 − 20 − 40 = 70 or 130 − 70 − 40 = 20 and 130 − 20 − 70 = 40.

Name:

Magic machines

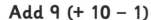

These two-step magic machines will help you to add or subtract numbers by using the nearest multiple of 10 and then adjusting.

The rule the machine follows is written over the top of the box and in each case the first one has been done for you to show you how the machine operates.

Add 9 (+ 10 – 1)

In	Middle	Out
16	26	25
24		
62		
89		
114		
153		

Subtract 9 (– 10 + 1)

In	Middle	Out
14	4	5
37		
73		
92		
125		
164		

Add 99 (+ 100 – 1)

In	Middle	Out
74	174	173
98		
123		
257		
406		
629		

Subtract 99 (– 100 + 1)

In	Middle	Out
161	61	62
193		
248		
416		
677		
821		

Dear Helper,

Talk with your child about how this kind of function machine works. Ask them about other function machines they might have used. What process did they carry out? Discuss how these particular machines involve a two-step process. After the number has been put in, the first process takes them into the middle section. The second process will produce the number that comes out of the machine.

Column addition

You will need: a sharp pencil and some centimetre squared paper.

Example

159 + 376

↓

```
  + 159
    376
   ────
     15
    120
    400
   ────
    535
```

Check

```
  + 159
    376
   ────
    535
    1 1
```

This is one method of column addition.

See how the sums have been written vertically.

Digits must go into the correct columns.

Squared paper will help you to do this.

The **units** are added first.

The **tens** are added next.

Finally the **hundreds** are added.

The answer can be checked using the short way.

- Find the answer to these addition questions using the method shown above. Then use the short way to check your answers. Work on the squared paper.

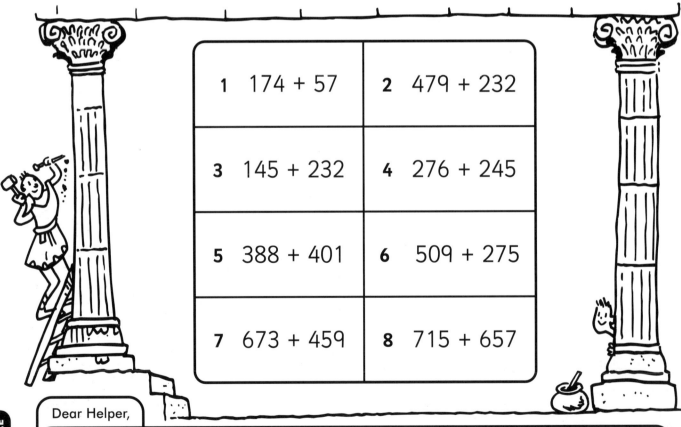

1 174 + 57	**2** 479 + 232
3 145 + 232	**4** 276 + 245
5 388 + 401	**6** 509 + 275
7 673 + 459	**8** 715 + 657

Dear Helper,

Talk through the method of column addition shown above with your child. Ask them to explain each of the stages to you to show that they understand. Check that they know how to set down the questions vertically, using the squared paper to help them keep the digits in the correct columns. Check that they add the units first, then the tens and finally the hundreds. Also ask them to describe the short method that they will use to check their answers at the end.

Column subtraction

You will need: a sharp pencil and some centimetre squared paper.

Example

475 − 87

$$475 = 400 + 70 + 5$$
$$-\ 87\ - \qquad 80 + 7$$
$$= 400 + 60 + 15$$
$$-\qquad\quad 80 + 7$$
$$= 300 + 160 + 15$$
$$-\qquad\quad 80 + 7$$
$$= 388 \quad 300 + 80 + 8$$

This is one method of column subtraction using **decomposition**.

475 is four hundred, seventy and five minus 87 which is eighty and seven.

We can exchange one 10 from the 70 leaving 60 and add it to the 5 to make 15.

We can exchange one hundred from the 400 leaving 300 and add it to the 60 to make 160.

When we subtract each column the result is 388.

- Find the answers to these subtraction questions using the decomposition method shown above. Work on the squared paper.

- Use the inverse operation (addition) to check your answer.
  For example: 475 − 87 = 388, so 388 + 87 = 475.

1 257 − 98	**2** 514 − 76
3 632 − 59	**4** 754 − 86
5 Subtract 56 from 497.	**6** Find the difference between 239 and 872.

Dear Helper,

Talk through the method of column subtraction shown in the example with your child. Ask them to explain each of the stages to you to show that they understand. Check that they know how to set down the questions using the squared paper to help them keep the digits in the correct columns. Also ask them to describe how the inverse operation (addition) can be used to check their answers at the end.

School stock

These are the costs of items needed for the school stock cupboard.

| lined exercise book 99p | notepad 38p | pencil 25p | tape measure 65p | ruler 32p |

| plain exercise book 89p | stapler £1.05 | sharpener 45p | rubber 30p | crayon 79p |

1 Find the cost of:

   a a ruler and notepad. ⬚

   b a lined book and a plain book. ⬚

   c a notepad and a crayon. ⬚

   d a pencil, a tape measure and a stapler. ⬚

2 How much change from £5.00 would you have after buying:

   a a stapler? ⬚

   b a crayon and a pencil? ⬚

   c a notepad and a ruler? ⬚

   d a tape measure and a sharpener? ⬚

   e 12 pencils? ⬚

   f both exercise books? ⬚

3 Some paint has been spilled in the stock cupboard on some of these amounts the teacher has been working out. Investigate, and then fill in, the missing numbers.

a
```
  pencil
+ 25p
  rubber
```

```
  55p
```

b
```
  sharpener   45p
  crayon    + 79p
```

c
```
  book
+
  notepad   38p
  137p
```

d
```
  ruler   32p
−
  pencil
  7p
```
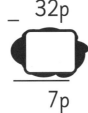

e
```
  lined exercise book
+
  sharpener   45p
  54p
```

f
```
  stapler      £1.05
  tape       −  65p
  measure
```
 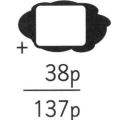

Dear Helper,

Go through the questions with your child to make sure they are able to write down the amounts vertically before carrying out an addition or subtraction. Your child may find working with sets of coins useful, if you have them available. Make sure that their answers are either written in pence or labelled with the £ sign. Investigate and discuss possible solutions to the final 'blot' questions.

Name:

Holiday travel

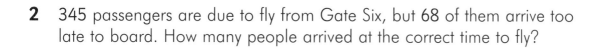

Many people fly from the local airport to go on their holidays.

- Help the airport workers to solve these problems.

- Underneath each question write down the numbers and the operations you use, then calculate the answer, recording how you work it out.

1 217 passengers fly from Gate One and 86 passengers from Gate Two. How many passengers fly altogether?

_____

2 345 passengers are due to fly from Gate Six, but 68 of them arrive too late to board. How many people arrived at the correct time to fly?

_____

3 In one day, 79 passengers travel to Spain, 105 travel to Italy and 236 fly to Greece. How many passengers travel altogether?

_____

4 There are 358 people waiting at Gate Seven and 179 fewer waiting at Gate Ten. How many are there at Gate Ten?

_____

5 359 passengers fly from Liverpool to Birmingham. At Birmingham, 109 of them get off, but 54 others get on. How many does the plane carry now?

_____

6 A plane flies most days during June, July and August, but it loses 17 days during these months because of routine engine maintenance. How many days does it fly?

_____

7 Of the 300 suitcases on a plane, one-half of them are grey and one-quarter of them are brown. The rest are blue. How many suitcases are blue?

_____

8 Flight One carries 125 adults and 29 children. Flight Two has 246 adults and 17 children. Flight Three has 198 adults and 51 children. How many adults fly on the three flights? How many children fly? How many people fly altogether?

Dear Helper,

Talk through the problems with your child and discuss any words that they do not understand. Discuss the real-life setting for these problems. Ask them to explain what type of operations (+, −, ÷ or ×) are needed to solve the problems and how they will set down the numbers in order to calculate the solutions. Carefully check those problems that involve a two-step operation.

Matching measures

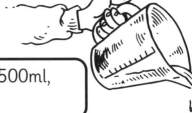

Remember   1000ml = 1 litre, $\frac{1}{4}$l = 250ml, $\frac{1}{2}$l = 500ml,
$\frac{3}{4}$l = 750ml and $\frac{1}{10}$l = 100ml.

1 Write these capacity measurements in at least one other way.

a 5l = _____ ml

b $3\frac{1}{2}$l = _____ ml

c $6\frac{1}{4}$l = _____ ml

d $9\frac{3}{4}$l = _____ ml

e $4\frac{1}{10}$l = _____ ml

f 2 500ml = _____ l

g 7 000ml = _____ l

h 3 750ml = _____ l

i 8 100ml = _____ l

j 12 250ml = _____ l

2 Fill in the missing amounts.

a 1 200ml = _____ l _____ ml

b 5 400ml = _____ l _____ ml

c 3 245ml = _____ l _____ ml

d 6 005ml = _____ l _____ ml

e 4l 350ml = _____ ml

f 8l 615ml = _____ ml

g 7.4l = _____ ml

h 9.7l = _____ ml

3 Put these capacities in order of size, beginning with the
smallest. It may help to write each capacity in ml.

a $\frac{1}{2}$l, 300ml, 0.6l

_____

b 800ml, $\frac{3}{4}$l, 0.9l

_____

c $1\frac{1}{4}$l, 2 000ml, 1.9l

d 3.7l, $3\frac{3}{4}$l, 3 900ml

_____

e 5 450ml, 5.9l, $5\frac{1}{4}$l

_____

f $4\frac{1}{4}$l, $4\frac{1}{10}$l, $4\frac{1}{2}$l

_____

Dear Helper,

Discuss what your child understands by the word 'capacity' – the amount that something holds. Check that
they know the information given in the fact box at the top of the sheet, especially the fractional parts of a
litre. See if they are able to write metric capacity measurements in more than one way, converting litres into
millilitres and vice versa, as well as using decimal equivalents, for example: ¼l = 250ml = 0.25l.

Name:

Capacity quiz

You will need: a 1 litre plastic measuring jug or a 1 litre plastic bottle and at least six containers such as drinks cans, small plastic bottles, mugs and egg cups.

Make sure you have permission to do this task first.

- Mark the jug or bottle with the measurements shown in the diagram opposite, if they are not labelled already.

- With your Helper, carefully fill the small containers with water until they are full.

- Estimate how much each will hold. Is it nearest to $\frac{1}{4}$l, $\frac{1}{2}$l, $\frac{3}{4}$l or 1l?

- Use the jug or the bottle to measure the amounts.

- Record your results in the chart below.

Container	Estimate ($\frac{1}{4}$l, $\frac{1}{2}$l, $\frac{3}{4}$l, 1l)	Measure (nearest to $\frac{1}{4}$l, $\frac{1}{2}$l, $\frac{3}{4}$l, 1l)

Dear Helper,

Please make some form of measuring jug or bottle and a number of clean containers available for your child to carry out this task. It would be advisable to work on a waterproof surface, or better still a draining board, near a tap. Help your child to fill the containers carefully and then carry out the measuring process, so that their results are as accurate as possible.

Mirror image

- Working with a friend, take it in turns to colour the coded squares in these four shapes.

The colours used are r = red, b = blue, y = yellow and g = green. Squares that are not labelled should be left blank.

- Repeat the same coloured pattern on the other side of the line or lines that have been drawn.

You will be making mirror symmetry reflections of the shapes.

1 and 2 have one line of symmetry. 3 and 4 have two lines of symmetry.

1

		b					
		g	r				
y	y	g	r				
		g	r				
		b					

2

				r			r
				b	g	g	y
				b	r	r	y
				b	g	g	y
				r			r

3

b	y	r			
y	r	y			
r	y	b			

4

r	g	b			
r	g	b			
r	g	b			

- Use squared paper to make some of your own coloured patterns with mirror or reflective symmetry. Choose your own rectangular or square grid size.

Dear Helper,

These patterns are examples of reflective symmetry, sometimes called mirror symmetry. If necessary, help your child to count squares across and up so that they repeat the pattern exactly on the other side of the line or lines. It may also help to check the finished coloured pattern by placing a plastic mirror on the line, or lines, of symmetry to make sure that the reflection is correct.

Name:

Letter land

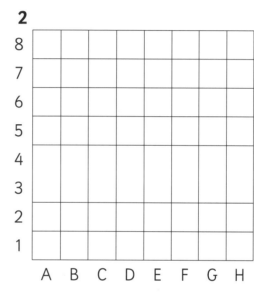

You will need: coloured pencils.

- Using the co-ordinate references, colour in the squares given for each grid 1, 2 and 3.

  What letter shapes have you formed?

- For grid 4 make up your own letter shape, but don't draw it on the grid.

- Write down the co-ordinates that need to be coloured to make this shape and see if your friend is able to find out which one it is.

1

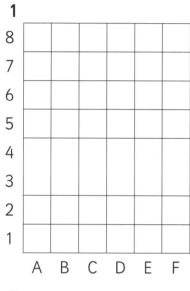

A1, A2,
A3, A4,
A5, A6,
A7, A8,
B8, C8,
D8, B8,
C8, D8
A5, B5,
C5.

2

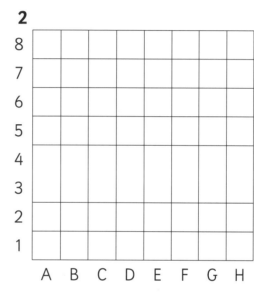

A2, B2,
C2, D2,
D3, D4,
D5, D6,
D7, E7,
F7, G7,
H7

3

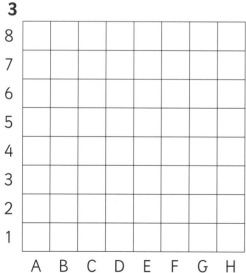

D1, D2,
D3, D4,
D5, C6,
B7, A8,
E6, F7,
G8.

4

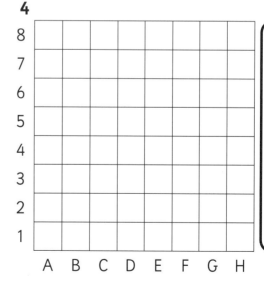

Colour
these
squares:

Dear Helper,

Point out to your child that the horizontal axis of each grid is labelled with letters and the vertical axis with numbers. Show them that, in this case, it is the spaces that have been labelled and not the lines. Make sure your child remembers that when plotting the positions, it will be the 'across' letters that will be read first, followed by the 'up' numbers.

Name: _____

Shape up

You will need: a ruler and a sharp pencil.

- For each grid first plot the points and then join them together in the order in which they are written. All three will make 2-D shapes.

- Make up your own 2-D shape for the blank grid.

- Write down the points and ask a friend to join them up to make the shape.

1

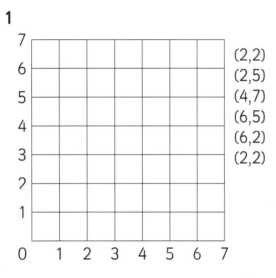

(2,2)
(2,5)
(4,7)
(6,5)
(6,2)
(2,2)

I have drawn a _____

2

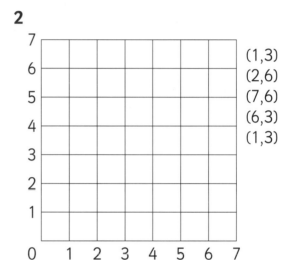

(1,3)
(2,6)
(7,6)
(6,3)
(1,3)

I have drawn a _____

3

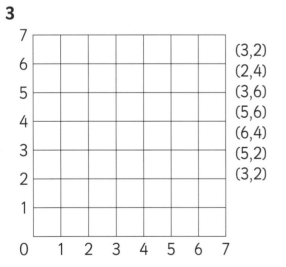

(3,2)
(2,4)
(3,6)
(5,6)
(6,4)
(5,2)
(3,2)

I have drawn a _____

4

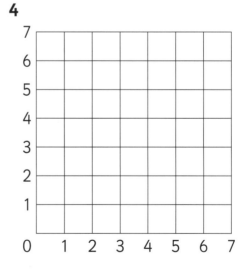

_____

Dear Helper,

Point out to your child that this time the lines, not the spaces, have been labelled and that numbers are used on both the axes. Remind them to plot the positions where the two lines cross and that horizontal (across) numbers are always written and plotted first. Stress that they should mark the points in the order given and that they should be joined with straight lines, using a ruler.

Name:

Right directions

You will need: a coloured pencil and your eight-point compass rose.

- Label each of the eight-point compass rose drawings that are given below.

- Mark each of the turns given on the drawing and then say what direction you would then have reached. C = clockwise and A = anticlockwise.

The first one has been done for you.

1

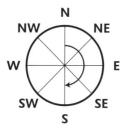

Start **N** $\frac{1}{2}$ turn C = _____**South**_____

2

Start **W** $\frac{1}{2}$ turn C = _____

3

Start **S** $\frac{3}{4}$ turn C = _____

4

Start **E** $\frac{3}{4}$ turn C = _____

5

Start **S** $\frac{1}{4}$ turn A = _____

6

Start **NE** $\frac{1}{4}$ turn C = _____

7

Start **SW** $\frac{1}{2}$ turn A = _____

8

Start **NW** $\frac{1}{4}$ turn A = _____

Dear Helper,

Talk through an eight-point compass rose with your child and encourage them to cover the example and label the drawings without help if they can. Also discuss the meaning of the words 'clockwise' and 'anticlockwise' and ask your child to demonstrate what they understand by a quarter of a turn, a half turn, three-quarters of a turn and a full turn.

Name: ·

Angle challenge

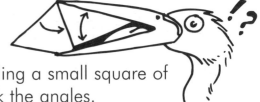

It may help to make a right-angle checker by folding a small square of paper like this. You could use it to help you check the angles.

Step 1

paper square

Step 2

fold in half

Step 3

right angle

fold in half again

All these angles are less than two right angles (180 degrees = 180°).

- Can you place them in order of size starting with the smallest?

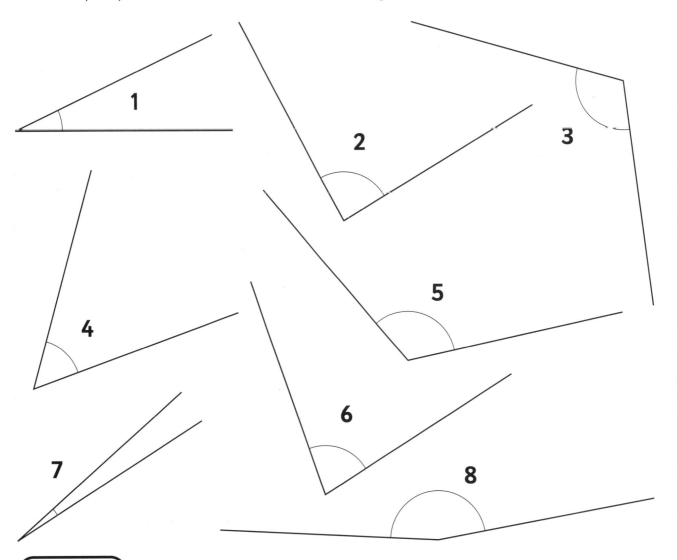

Dear Helper,

Remind your child that the word 'angle' means the amount of a turn. The bigger the turn, the larger the angle. Help them with the construction of the right-angle checker. A right angle measures 90°. Start by asking if they can identify the angles that are less than 90°, the angles that are about 90°, those that are bigger than 90° and those that are about 180°.

Multiple sort

- Look at each number in the left-hand column and decide whether it is a multiple of 2, 4, 5 or 10.

- Tick the columns when it is a multiple.

	Multiple of 2	Multiple of 4	Multiple of 5	Multiple of 10
1	—	—	—	—
2	✔	—	—	—
3	—	—	—	—
4	✔	✔	—	—
5				
6				
7				
8				
9				
10				
11				
12				
13				
14				
15				
16				
17				
18				
19				
20				
21				
22				
23				
24				
25				
26				
27				
28				
29				
30				

Dear Helper,

Your child has been learning to recognise multiples of numbers and this investigation will help them to sort multiples. Encourage them to look at each number and decide if it is divisible by 2, then by 4, 5 or 10. Remind them to use known facts, such as 'all multiples of 10 end with 0'. Your child may like to continue the investigation on another piece of paper, looking at numbers up to 50 or 100. Alternatively, they could look for multiples of other numbers such as 3.

Name:

The rule of three

The 3 times table has been written on the chart.

- Add together the digits in each product to get the digit total. The first ones have been done for you. Can you see the pattern?

- Continue the 3 times table up to 20 × 3, then add together the digits to see if the pattern is continued. You can use a 100 square to help you count on.

Multiply	Product	Add the digits	Total of the digits
1 × 3	= 3	3	3
2 × 3	= 6	6	6
3 × 3	= 9	9	9
4 × 3	= 12	1 + 2 =	3
5 × 3	= 5	1 + 5 =	6
6 × 3	= 18	1 + 8 =	9
7 × 3	= 21	2 + 1 =	
8 × 3	= 24	2 + 4 =	
9 × 3	= 27	2 + 7 =	
10 × 3	= 30	3 + 0 =	
11 × 3			
12 × 3			
13 × 3			
14 × 3			
15 × 5			
16 × 3			
17 × 3			
18 × 3			
19 × 3			
20 × 3			

- Write down what you have noticed on the back of this sheet.

- Investigate with some higher multiples of 3.

Dear Helper,

Please discuss this investigation with your child. If they have difficulty working out the multiples of 3, encourage 'counting on'. Your child may like to continue the investigation with much bigger multiples, in which case it may be helpful to allow the use of a calculator.

Is it a multiple?

- Try to do this exercise on your own as quickly as you can.

1 Circle the multiples of 5   12,   15,   23,   30,   34,   35,   41,   55.

2 Circle the multiples of 2   10,   13,   16,   23,   26,   30,   35,   40.

3 Circle the multiples of 10   10,   22,   30,   34,   50,   60,   65,   90.

4 Circle the multiples of 4   12,   20,   22,   30,   32,   38,   40,   44.

5 Circle the multiples of 3   6,   10,   14,   18,   21,   28,   33,   40.

6 Circle the numbers which are multiples of **both** 3 and 4.

8,   12,   15,   18,   24,   30,   36,   40,   60.

7 Circle the numbers which are multiples of **both** 2 and 10.

8,   12,   15,   20,   22,   24,   30,   38,   40.

8 Circle the numbers which are multiples of **both** 5 and 10.

9,   10,   12,   15,   30,   36,   42,   45,   50.

Dear Helper,

This activity is to help your child identify simple multiples of numbers quickly. Remind them to check each of the numbers carefully, using the tests for recognising multiples that have been taught in school.

Name: _____

Tables square

You will need: a stop-watch or watch with a seconds hand.

• Complete this tables square as quickly as you can.

Ask your Helper to time you.

×	1	2	3	4	5	6	7	8	9	10
1										
2										
3										
4										
5										
6										
7										
8										
9										
10										

• Use the tables square to help you complete these multiplications:

2 × 7 =   4 × 6 =   9 × 9 =   8 × 7 =

4 × 9 =   7 × 6 =   8 × 3 =   9 × 3 =

4 × 8 =   2 × 6 =   6 × 5 =   8 × 10 =

6 × 6 =   8 × 8 =   5 × 5 =   10 × 10 =

Time taken _____ **mins** _____ **secs**

Dear Helper,

Your child has been learning different mental strategies, such as doubling and counting on. They should use these strategies to complete the multiplication square as quickly as possible. Your child may then use it to find the solutions to the multiplications. You may like to give your child some more examples to complete using the 'square'.

Double it!

- Match each number with its double!

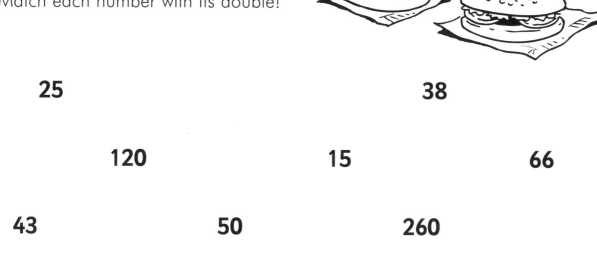

25   38

      120   15   66

43   50   260

      19   33   240

520   30   86

- Now play a challenge game with your Helper.

- Ask your Helper to write down a number and say either 'double' or 'halve'!
  Then you write down the matching number.

- See how many you can get right. Can you get five right in a row?

Dear Helper,

Your child has been practising doubling and halving numbers and using the facts they already know to help double larger numbers. Give your child a series of numbers to either double or halve, such as 20, 43, 19 or 36. You can vary the numbers to suit your child's ability.

Name:

Times it!

Here is a method for multiplication: 46 × 8

Approximating gives 45 × 10 = 450

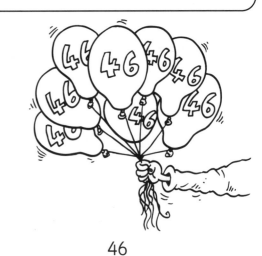

8 × 6 = 48
Write the **8** units under the
other units and 'carry' the **4** tens

```
    46
×    8
_____
     8
_____
     4
```

8 × 40 = 320, add on the 40 or 4 tens to give 360.
Add this to the 8 in the units column to give the answer

```
    46
×    8
_____
   368
_____
     4
```

- Do these multiplications in the same way, remember to approximate first.

1
```
    57
×    7
_____

_____
```

2
```
    49
×    6
_____

_____
```

3
```
    63
×    8
_____

_____
```

4
```
    94
×    4
_____

_____
```

5
```
    77
×    7
_____

_____
```

6
```
    84
×    5
_____

_____
```

Dear Helper,

You could time your child to see how quickly they can do these calculations. Talk about the need for approximation first and then using this to check whether the final answer 'looks right'. If your child needs help, talk them through the stages shown in the example.

Name:

Divide it!

Here is a method for dividing 96 by 8.

First we approximate. We know that 8 is close to 9, and 10 × 9 = 90, so our answer is going to be a bit higher than 10, so a good approximation may be 11.

We know that 10 × 8 is 80:

Take the 80 away from 96 leaving 16

We know that 2 × 8 is 16:

Take 16 away from 16:

Thus: 12 × 8 = 96 and 96 ÷ 8 = 12

$$\begin{array}{r} 8\overline{)96} \\ -80 \\ \hline 16 \\ -16 \\ \hline 0 \end{array}$$

10 × 8

2 × 8

10 + 2 = 12

Do these divisions using the same method:

1 75 ÷ 5	**2** 66 ÷ 3
3 92 ÷ 4	**4** 72 ÷ 3
5 95 ÷ 5	**6** 48 ÷ 4

Dear Helper,

This is a practice exercise to help your child to gain confidence with column division. Let them try to complete the exercise on their own, but if help is needed talk through each step as shown in the example. You may like to give some further practice examples such as 78 ÷ 3, 85 ÷ 5, 56 ÷ 4.

Name:

Remainders

Here are some divisions which may have a remainder in the answer.

- Work out each division carefully using the column division method that you have been practising. Remember to approximate first!

1 99 ÷ 7	**2** 73 ÷ 6	**3** 82 ÷ 5
4 90 ÷ 4	**5** 78 ÷ 6	**6** 67 ÷ 4

- You can try these extra problems on the back of this worksheet.

Extension problems:

a 5 children share a box of 72 sweets. How many do they have each and are there any left? _____

b The class has 84 crayons. How many can each of the 6 groups have? _____

Will there be any left? _____

Dear Helper,

Your child has been learning about remainders after division. Encourage them to do these calculations using 'column division', pointing out that there may be remainders with some of the questions. You may like to help your child try the two extension problems.

Name:

Party shopping game

You will need: a six-sided dice (or game cards) and some extra paper to keep a record of your purchases.

Rules for the game:

The aim of the game is to 'buy' everything on the shopping list.

- Taking turns, throw the dice and look on the 'Money list' to see how much money you have to spend.

- Look at the shopping list and the price list and decide what you can buy. Work out the cost. You may save the money you have left and add it to the amount for your next turn.

The winner is the person who buys everything on the 'Shopping list' first.

Price List	
Paper cups	5p
Paper plates	10p
Individual cakes	20p
Large drink	50p
Packet of crisps	8p
Sausage rolls	9p

Shopping List
6 paper cups
6 paper plates
6 individual cakes
2 large bottles of pop
6 packets of crisps
6 sausage rolls

Money		
1	–	50p
2	–	£1
3	–	60p
4	–	45p
5	–	90p
6	–	75p

Start recording your purchases here.

Dear Helper,

Please play this game with your child. You may like to make some game cards showing the amounts of money to choose from, rather than throwing a dice. The activity will help them to gain an understanding of remainders after division in a real-life situation with money.

Name:

Fraction decimal match

- Label this grid in fractions and decimals. Some have been done for you.

$\frac{1}{10}$									

0   0.5   0.8

- Match each fraction to its decimal equivalent. You can use the grid to help.

Left column (fractions): $\frac{1}{4}$, $\frac{3}{10}$, $\frac{1}{10}$, $1\frac{1}{2}$, $1\frac{1}{10}$, $\frac{7}{10}$, $3\frac{1}{2}$, $10\frac{1}{10}$, $\frac{3}{4}$, $\frac{9}{10}$

Right column (decimals): 10.1, 0.75, 0.3, 3.5, 0.25, 0.9, 0.1, 1.5, 1.1, 0.7

Dear Helper,

This exercise is to help your child match fractions to decimals. Encourage your child to look at the grid if they need help, pointing out that, for example, ³⁄₁₀ is equivalent to 0.3. You may like to give some further examples using mixed fractions such as 3 ³⁄₁₀.

Name:

Highest/lowest

You will need: a set of 0–9 digit cards and a decimal point card.

Rules:

- Players take turns.

- Lay the shuffled digit cards face down.

- First player selects three cards and the decimal point.

- They then have to use the four cards to make the highest or the lowest number possible, declaring 'highest' or 'lowest'. All four cards must be used.

- If the other player can make a higher (or lower) number then they win.

Record your numbers here:

Highest		Lowest	
Player 1	**Player 2**	**Player 1**	**Player 2**

Dear Helper,

Play 'Highest/lowest' with your child. The aim is to position the digits and the decimal point correctly to achieve the target of a high or low number. Talk to your child about the different numbers that can be made with each set of cards. For example, 1, 2, 3, . could give 32.1 as the highest, but 1.23 as the lowest, with a range of other numbers in between such as 2.31, 23.1 and 31.2.

Decimal hunt

Decimals are used in lots of measurements and quantities.

- Look at items around your home and make a list of any quantities or prices that you find on labels.

- See how many different quantities you can find and put them on the list.

- Can you convert the quantities and prices on the labels into decimals?

Object	Decimal quantity and details
Kitchen foil 20m long by 300mm wide 95p	20m × 0.3m £0.95

Name:

Money adds

1 £3.45
  + £6.24

  _____

2 £5.99
  + £3.25

  _____

3 £4.05
  + £8.58

  _____

4 £4.95
  + £2.50

  _____

5   75p
  + £2.50

  _____

6 £2.45
  + £9.75

  _____

7 Add together £3.25 and £10.05.

8 Add together £6.70, 95p, and £3.05.

9 Calculate £4.30 + 75p + £2.08 + 5p.

Dear Helper,

This is an exercise to help your child practise using column addition of money. Your child should be able to do this without help. If necessary make sure that they position the amounts correctly so that the decimal points are 'in line', especially with the last three examples. You may like to give some more, similar examples for your child to practise.

Name:

Shopping check!

You will need: some supermarket till receipts.

- Carefully cut a supermarket till receipt into at least three sections.

- Using one section at a time, list the amounts and total them.

Section 1	Section 2	Section 3
Total	**Total**	**Total**

- Total all three sections.   **Grand total**

- Is your total the same as the total on the bill? _____

Dear Helper,

Help your child find a suitable till receipt to check, ideally with at least ten items. Cut the receipt into three and keep the final, 'grand' total separate. Your child should list the amounts on each section of the bill and total them. They can check their accuracy by adding the three amounts together and comparing their grand total with the receipt total. You can vary this activity to suit your child by providing longer or shorter receipts.

Name:

TV times

You will need: a TV programme guide from a newspaper or magazine.

- Plan a **Children's TV Bonanza**. Imagine you are a TV planner and you are going to present a special day's viewing for children in the school holidays.

- Look at the different TV programmes and choose a selection.

- Complete the TV planner below for the 'Children's Bonanza'.

TV Programmes	Length of programme	Start time
Cartoon Capers	20mins	9.00a.m.

Dear Helper,

Please help your child choose some programmes for their selection. Help them to check the times and length of programmes and fit them into an overall plan. The activity will help your child to understand simple timetables, as well as calculations involving time.

Name:

Carroll sort

- Make a list of some of your family and friends and note down their hair and eye colours.

Name	Eye colour	Hair colour

- Can you complete the Carroll Diagram?

- Decide which colour eyes and which colour hair to put on the diagram and then add the names from your list.

	_____ eyes	Not _____ eyes
_____ hair		
Not _____ hair		

Dear Helper,

Your child has been learning about sorting diagrams like the Carroll diagram on this worksheet. Help them to collect information about the hair and eye colours of several family members and friends. They should decide which words to use to describe hair and eye colour in the chart and then put the people's names in the correct 'box' on the diagram.

Which circle?

Name:

A B C D E F G H I J K L M N O P Q R S T U V W X Y Z
a b c d e f g h i j k l m n o p q r s t u v w x y z

- Look carefully at each of the capital letters, can you spot which ones have reflective symmetry?

- Which capital letters look the same as their lower case version?

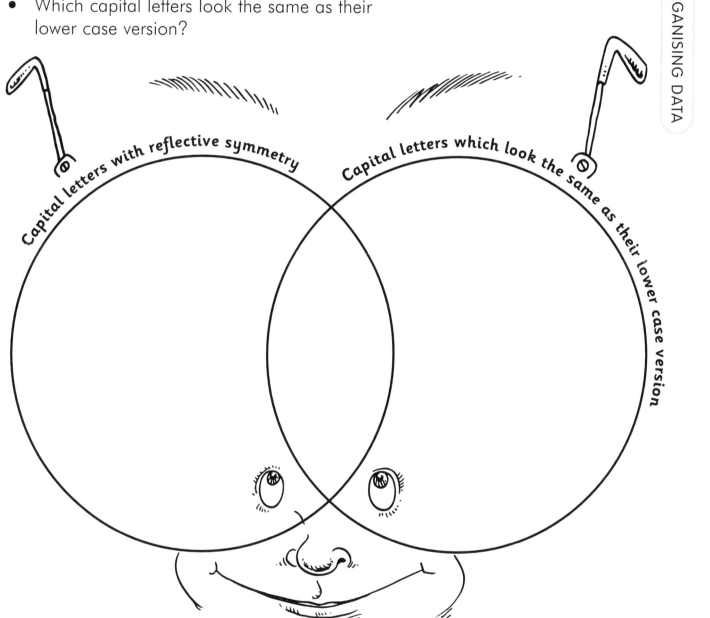

Capital letters with reflective symmetry

Capital letters which look the same as their lower case version

- Can you complete the Venn diagram?

Dear Helper,

Please help your child to look carefully at each letter and decide whether it is symmetrical and whether the upper and lower case versions are the same. If the letter meets both criteria (such as O), it should be written in the overlap of the two circles. If it does not meet either criteria (such as G), it should be written outside the circles.

Name:

Houses and homes

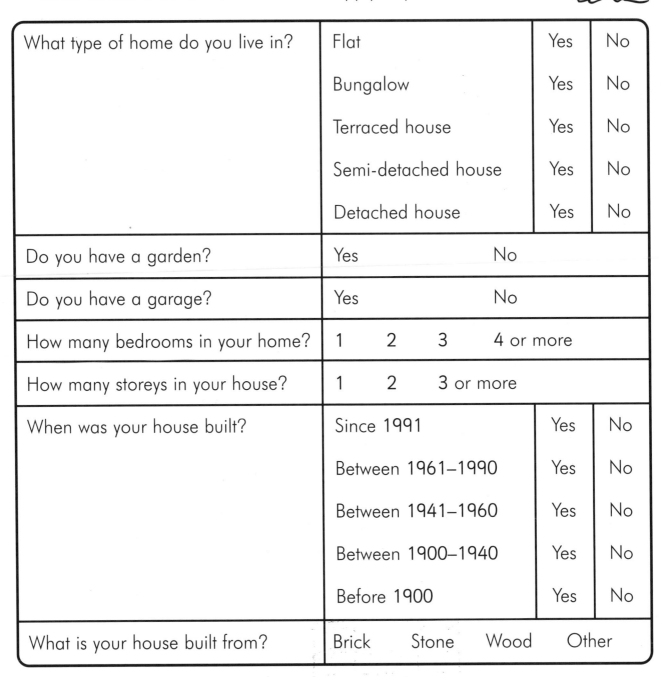

- Complete this questionnaire about your home.

- When you get back to school you can input the data into the class database on the computer.

- Draw circles around the answers that apply to your home.

What type of home do you live in?	Flat	Yes	No
	Bungalow	Yes	No
	Terraced house	Yes	No
	Semi-detached house	Yes	No
	Detached house	Yes	No
Do you have a garden?	Yes No		
Do you have a garage?	Yes No		
How many bedrooms in your home?	1 2 3 4 or more		
How many storeys in your house?	1 2 3 or more		
When was your house built?	Since 1991	Yes	No
	Between 1961–1990	Yes	No
	Between 1941–1960	Yes	No
	Between 1900–1940	Yes	No
	Before 1900	Yes	No
What is your house built from?	Brick Stone Wood Other		

- Can you think of another question we could ask about people's homes?

Dear Helper,

The children in your child's class are collecting data for a computer database. Please help your child complete the questionnaire so that they can input the data as accurately as possible into the computer.